CATASTROPHIC COLLAPSES

FROM VISION TO VULNERABILITY

A Global Timeline of Human, Technical, and Moral Failures

AJOY MAJUMDAR

This book is a carefully assembled account of real-world incidents, based on publicly available and credible sources. The author makes no claim of firsthand involvement in the events described. Interpretations and perspectives offered are those of the author.

This book is intended for informational and educational purposes only. While every effort has been made to ensure accuracy at the time of writing, readers are encouraged to consult cited sources for further study or verification.

The content herein is not intended to serve as case-specific technical guidance or as a substitute for formal professional consultation.

ISBN: 978-1-999716-00-6

First Edition
Published in Canada

Table of Contents

Acknowledgements

Writing *Catastrophic Collapses* has been a journey through blueprints, failures, and the human stories behind them. I am grateful to the many individuals, including engineers, researchers, and chroniclers, whose documentation and analysis of past failures have made this book possible.

Special thanks to the global engineering and construction community, both past and present, for their continued commitment to learning from past mistakes and shaping a safer, more resilient world.

To my family and friends, I deeply appreciate your quiet support, your patience, and your steady encouragement during long edits and late nights.

And finally, to the readers: may these stories not only inform, but also inspire vigilance, accountability, and the pursuit of excellence in everything you build.

Preface

Engineering has always walked a fine line between ambition and consequence. The structures we create—bridges, dams, tunnels, towers—stand as powerful symbols of human ingenuity, but also as reminders of how fragile our achievements can be when something fails.

This book brings together engineering-related disasters from around the world. Some were caused by design flaws. Others resulted from negligence, oversight, or a refusal to act on known risks. In many instances, the engineering itself held up, but decisions made around it—managerial, political, or procedural—set the stage for failure. These events show that no structure operates in isolation. Design, materials, human judgment, maintenance, and accountability are all tightly linked.

The stories are arranged chronologically to trace how such failures have recurred across time. Despite major technological advances, many of the same weaknesses continue to reappear. Spanning different regions and industries, each case left a scar—and each has something to teach us.

This book aims to preserve those lessons. It does not seek to point fingers, but rather to explore how and why these failures happened. In doing so, it hopes to encourage deeper awareness, responsibility, and resilience in how we shape the world around us.

AM

Introduction

Engineering, at its core, transforms the world. It builds bridges that span rivers, tunnels that pierce through mountains, and systems that power cities and connect continents. But when precision falters, when oversight lapses, when ambition overrides caution, or when corners are cut, the consequences can be catastrophic.

This book, *Catastrophic Collapses*, chronicles 50 major engineering-related disasters that left lasting impacts around the world. From collapsed bridges and crumbled towers to dam breaches, industrial explosions, and chemical catastrophes, each entry uncovers more than just broken concrete or scorched steel. These stories also reveal the human decisions, including miscalculations, ignored warnings, flawed designs, and systemic neglect, that paved the path to failure.

While the book focuses on fifty major events, they represent only a fraction of the countless engineering-related disasters that have occurred throughout modern history. To acknowledge this broader context, many chapters include references to similar incidents in other regions. These connections serve as reminders that such tragedies are often part of recurring patterns rather than isolated anomalies. Whether it is a bridge collapse in Genoa that echoes earlier failures in Florida and Bangladesh, or an industrial disaster in Bhopal that mirrors systemic lapses seen in Texas City or Kathmandu, these links underscore the global and often repeated nature of engineering oversight gone wrong.

Spanning continents and decades, the selected incidents range from the infamous, like Chernobyl and Bhopal, to lesser-known but equally devastating cases, such as construction failures in Nairobi, Phnom Penh, or Singapore. The goal is not only to recount what went wrong, but also to examine why it happened and what can be learned from it, especially in a world increasingly dependent on complex and interconnected infrastructure.

The information presented here has been carefully compiled from publicly available sources, including investigative reports, news archives, historical records, and expert analyses. References and cross-links are included at the end of each entry for readers who wish to explore further.

Whether you are an engineer, policymaker, educator, student, or simply someone fascinated by the invisible threads that hold modern life together, this book offers enduring lessons in humility, accountability, and the critical importance of engineering integrity.

Let these failures remind us that behind every blueprint, process flow, or control panel lies a duty. The goal is not just to complete the job, but to protect the lives that depend on it. In engineering, responsibility is not optional. It is structural.

Part I:
The Early Lessons
(Pre-1950)

1.
The Ålesund Fire-Watch Tower Collapse (Norway, 1904):
A Warning Lost in Flames

On the icy evening of January 23, 1904, a catastrophic fire swept through the picturesque town of Ålesund on Norway's western coast. Fueled by wooden buildings, fierce winds, and frigid temperatures, the blaze devastated over 800 structures and displaced more than 10,000 residents. But amid the destruction, one collapse stood out—not of a home or warehouse, but of the town's fire-watch tower.

The fire-watch tower had been erected in the late 19th century to serve as an elevated post for spotting smoke and coordinating emergency response. Perched on a steep hillside, it had stood as a quiet sentinel over Ålesund—until the night it failed its most critical duty. As flames raced through the town, the tower collapsed, not from fire directly, but from years of structural neglect.

Investigations later revealed that the tower's foundation had suffered repeated frost heaving over winters, compromising its stability. Cracks in its stone base had gone unaddressed, and the iron reinforcements had corroded over time. The collapse occurred just as the fire reached the hillside, and falling debris and blocked paths hindered fire brigades during the crucial early hours.

Though less famous than other engineering failures, the Ålesund tower incident highlighted the dangers of overlooking maintenance in public safety infrastructure. The tragedy led to municipal reforms and mandatory audits of civic structures. The rebuilt town, known today for its Art Nouveau architecture, rose from the ashes—but the silent fall of its watchtower remains a potent reminder.

The tower stood to protect the town from fire—but in the end, it had no defense against its own slow decay. Its fall was quiet, but it asked the loudest question of all: who watches the watchtower, when maintenance turns invisible?

Built to guard, it fell unguarded. The Ålesund tower's collapse was not from flame, but from forgetfulness—a subtle reminder that even symbols of safety must themselves be made safe.

Sources (*and related articles*):

- *Wikipedia.* "Ålesund fire." https://en.wikipedia.org/wiki/%C3%85lesund_fire
- *Nikel, David.* "The Story of the Ålesund Fire of 1904." *Life in Norway*, Jan 30, 2023. https://www.lifeinnorway.net/alesund-city-fire/
- *Karaliova, Larysa.* "Ålesund—The Art Nouveau City." *Scientific Research Publishing*, 2015. https://www.scirp.org/html/2-2140005_57562.htm
- *Ålesund Museum.* "The 1904 City Fire." *Utstilling – Bybrannen 1904.* https://www.aalesunds.museum.no/en/utstillinger/bybrannen-1904

- *Forbes*. "The Story Behind The Stunning Architecture Of Ålesund, Norway." Apr 15, 2022. https://www.forbes.com/sites/davidnikel/2022/04/15/the-story-behind-the-stunning-architecture-of-lesund-norway/

2. The Courrières Mine Disaster (France, 1906): *A Thousand Voices, Buried Still*

On the morning of March 10, 1906, the Courrières coal mine in northern France exhaled a plume of smoke, dust, and death. A colossal underground explosion ripped through the shafts, killing 1,099 miners in an instant. Many were boys no older than twelve. Entire families were obliterated, leaving the nearby villages cloaked in a grief as thick as the coal soot that lined their walls.

The blast traveled like a curse—igniting coal dust and methane gas in violent succession, amplified by a deadly cocktail of engineering oversights. Poor ventilation design, the absence of explosion-proof lighting, and flawed emergency exits turned the mine into a tomb. The explosion collapsed entire galleries and sealed escape routes, entombing hundreds alive.

What followed deepened the tragedy. In a decision still criticized over a century later, rescue operations were halted prematurely by the mining company, which feared risking more lives and—less nobly—its profits. Days later, 13 miners staggered out on their own, having survived underground for 20 days by navigating collapsed tunnels and drinking seepage water. Their emergence, miraculous yet damning, ignited national fury.

The disaster triggered widespread protests, labor strikes, and a reckoning with France's industrial conscience. It led to lasting reforms in mining safety across Europe—ushering in stricter ventilation protocols, worker protections, and emergency planning standards.

Yet the Courrières mine remains not just a cautionary tale, but a silent necropolis—where the earth once breathed fire, and never gave back all it took.

The mine did not collapse all at once—it failed in decisions, in design, in disregard. And when it finally spoke, it spoke in dust, fire, and a silence that lasted longer than rescue ever did. Decades later, in Benxihu (China) and again in Soma (Turkey), the silence returned—proof that deep underground, the same mistakes often wait patiently to be made again.

Sources (*and related articles*):

- *Wikipedia.* "Courrières mine disaster." https://en.wikipedia.org/wiki/Courri%C3%A8res_mine_disaster
- *Encyclopaedia Britannica*, "Courrières mining disaster." Available via Britannica. https://www.britannica.com/event/Courrieres-mining-disaster
- *History.com*, "Mine explosion kills 1,060 in France," Nov 13, 2009. https://www.history.com/this-day-in-history/march-10/mine-explosion-kills-1060-in-france

- *EHNE*, "European Perspectives on a Disaster: Courrières (1906)." https://ehne.fr/en/encyclopedia/themes/material-civilization/risks-and-security/european-perspectives-a-disaster-courri%C3%A8res-1906
- *SciSpace PDF*, "Reactions to the 1906 Courrières Colliery Mine Disaster." https://scispace.com/pdf/disaster-and-discourse-reactions-to-the-1906-courrieres-1yyonvnph1.pdf

10

3.
The Quebec Bridge Disaster (Canada, 1907 & 1916):
The Bridge That Collapsed Twice

The saga of the Quebec Bridge stands as one of the most damning cautionary tales in civil engineering—a colossal structure that failed not once, but twice, claiming nearly 90 lives. Intended to be the longest cantilever bridge in the world, the project began in 1900 to connect Quebec City with the south shore of the St. Lawrence River, serving Canada's rapidly expanding railway system.

The bridge was an ambitious endeavor led by the Phoenix Bridge Company and overseen by consulting engineer Theodore Cooper. But hubris and poor communication would soon spell disaster. Junior engineers raised concerns about bending stresses and load distribution as early as 1906. Cooper, working remotely from New York and reluctant to face scrutiny, dismissed the warnings.

On August 29, 1907, with the bridge still under construction, the south cantilever arm suddenly buckled and collapsed into the river. Seventy-five workers were killed. A royal commission later determined that the structure had been grossly underdesigned, and that load calculations failed to account for the true weight of the materials.

A second attempt began in 1913, this time with more rigorous oversight. But on September 11, 1916, as crews attempted to lift the central span into place, the hoisting system failed. The 5,100-ton span plunged into the river, killing 13 more men. It wasn't until 1917 that the bridge was finally completed.

The Quebec Bridge remains the longest cantilever bridge in the world at 987 meters. But it is equally remembered as the graveyard of misjudged math, bureaucratic ego, and avoidable human loss. It catalyzed reforms in engineering education, peer review, and construction oversight across North America.

Québec Bridge became more than a failed structure; it became a case study in the dangers of unchecked confidence and the cost of ignoring caution. Twice it collapsed—first under the weight of flawed calculations, then again through rushed execution—exposing systemic gaps in communication, verification, and professional responsibility. It took three attempts, dozens of lives, and a transformation in engineering culture before the bridge finally stood. Today, it endures not simply as a span of steel across the St. Lawrence, but as a permanent fixture in design handbooks, ethics classrooms, and every conversation that begins with the question: "What did we miss?"

Sources (*and related articles*):

- *Wikipedia*. "Quebec Bridge." https://en.wikipedia.org/wiki/Quebec_Bridge
- *American Society of Civil Engineers (ASCE). The Collapse of the Quebec Bridge* (PDF Case Study). https://collaborate.asce.org/HigherLogic/System/DownloadDocumentFile.ashx?DocumentFileKey=3d06bb6b-be0d-f9ce-0ffd-fc5160d673b9

- *CSCE Historic Summary*: https://legacy.csce.ca/en/historic-site/the-quebec-bridge/
- *Scribd Case Summary*: https://www.scribd.com/document/486774863/The-Quebec-Bridge-Collapse-Final-Paper

4.
Boston Molasses Disaster (USA, 1919):
The Sweet Deluge That Soured a City

On a frigid afternoon of January 15, 1919, the industrial heartbeat of Boston's North End skipped—then burst. A towering steel tank, groaning under 2.3 million gallons of molasses, suddenly ruptured. What followed was surreal and deadly: a brown wave 25 feet high thundered through the streets at 35 miles per hour, engulfing everything in its path—horses, freight cars, buildings, and people.

Twenty-one lives were lost. Another 150 were injured. Survivors spoke of being knocked unconscious by the force, or stuck and nearly drowned in the viscous flood. The air reportedly smelled sweet for weeks. But the aftermath was anything but.

The culprit wasn't nature—it was negligence. The United States Industrial Alcohol Company had rushed the tank into operation during wartime demand. It was never pressure-tested. Its rivets were carelessly installed, the steel walls too thin and prone to brittle fracture. Local residents had long reported ominous groans and leaks—molasses oozed so regularly that children would collect it with sticks. But the company did nothing.

In the wake of the disaster, public outrage fermented. Investigations revealed systemic failures not just in engineering, but in regulatory oversight. The incident became a cornerstone case in American industrial law, helping to establish accountability standards, mandatory testing, and formal inspections for large pressure vessels.

Today, the "Great Molasses Flood" sounds almost comedic in name—but its legacy is a sobering reminder of what happens when sweet profits outweigh structural prudence.

It wasn't the molasses that killed—it was the rush to fill a tank unfit for what it was meant to hold. The Boston disaster became a lasting reminder that not all failures are mechanical. Some begin with the belief that no one will notice… until the whole thing bursts.

Sources (*and related articles*):

- *Wikipedia.* "Great Molasses Flood." https://en.wikipedia.org/wiki/Great_Molasses_Flood
- *History.com Editors.* "The Great Molasses Flood of 1919." Last updated May 27, 2025. https://www.history.com/articles/the-great-molasses-flood-of-1919
- *Sohn, Emily.* "Why the Great Molasses Flood Was So Deadly." *History.com,* Jan 15, 2019. https://www.history.com/articles/great-molasses-flood-science
- *Puleo, Stephen.* Dark Tide: The Great Boston Molasses Flood of 1919. *Beacon Press, 2003.*

- "Boston Molasses Flood: Centennial Commemoration." *Boston.gov*, Jan 15, 2023. https://www.boston.gov/news/100-years-ago-today-molasses-crashes-through-bostons-north-end

5.
Knickerbocker Theatre Collapse (USA, 1922):
A Still Night, Then Silence

On the frigid evening of January 28, 1922, as a blizzard blanketed Washington, D.C., hundreds of theatergoers sought refuge inside the ornate Knickerbocker Theatre, eager for warmth and a silent film. But just as the reels flickered across the screen, the building itself fell silent. Then it fell completely.

With over 3 feet of snow weighing down its flat, poorly supported roof, the structure suddenly gave way—crushing patrons beneath tons of timber, steel, and plaster. Ninety-eight people were killed, and over 130 injured, making it one of the deadliest building collapses in U.S. history.

The disaster wasn't simply an act of nature—it was a failure of design, judgment, and oversight. The theater's roof relied on long-span steel trusses that lacked proper bracing. There was no structural redundancy, and the building codes of the time didn't adequately address snow load calculations in regions where snowfall was rare but not impossible. Witnesses recalled the ominous creaks before the collapse—signs that the roof was speaking, if anyone had been listening.

The aftermath triggered a sweeping overhaul of building codes across the United States, particularly for public assembly structures. Requirements for snow load considerations, truss inspections, and emergency response planning were strengthened.

The Knickerbocker's fall became more than a cautionary tale—it became a foundational lesson in architectural humility and the unpredictable burden of forgotten forces.

The Knickerbocker Theatre didn't just fall—it restructured how engineers and architects think about live loads, especially in public spaces. Its legacy lives not in concrete or steel, but in revised codes, reinforced roofs, and the persistent reminders that came decades later—from Hartford (CT, USA) in 1978 to Katowice (Poland) in 2006—that snow may fall quietly, but never without consequence.

Sources (*and related articles*):

- *Wikipedia.* "Knickerbocker Theatre (Washington, D.C.)." https://en.wikipedia.org/wiki/Knickerbocker_Theatre_(Washington,_D.C.)
- *Smithsonian Magazine.* "Nearly 100 Washingtonians Died When a Theater Collapsed After One of the Largest Snowfalls Ever to Hit D.C." (2022). Retrieved from: https://www.smithsonianmag.com/smart-news/nearly-100-washingtonians-died-when-theater-collapsed-one-largest-snowfalls-ever-hit-dc-180985871/

- *Smithsonian Magazine*. "When a Winter Storm Triggered One of the Deadliest Disasters in D.C. History." *Smithsonian Magazine* (January 26, 2022). Retrieved from: https://www.smithsonianmag.com/history/when-a-winter-storm-triggered-one-of-the-deadliest-disasters-in-washington-dc-history-180979446/
- *The Guardian*. "Weatherwatch: the tragic Knickerbocker snowstorm of 1922." *The Guardian* (January 27, 2022). Retrieved from: https://www.theguardian.com/news/2022/jan/27/weatherwatch-tragic-knickerbocker-theatre-snowstorm-washington-1922
- *DCist*. "Knickerbocker Collapse Killed Dozens 100 Years Ago." *DCist* (January 28, 2022). Retrieved from: https://dcist.com/story/22/01/28/dc-biggest-disaster-100-years-ago-knickerbocker-theater/

6.
St. Francis Dam Collapse (USA, 1928):
The Engineer Who Couldn't Sleep

In the early hours of March 13, 1928, a wall of water roared through the narrow canyons of northern Los Angeles County, obliterating everything in its path. Trees, homes, cattle, cars, and entire communities were swept away. When dawn broke, the St. Francis Dam was gone—and so were more than 400 lives.

Built to support the growing demands of Los Angeles' water supply, the dam had been completed in 1926 under the direction of William Mulholland, the famed engineer behind the Los Angeles Aqueduct. But Mulholland's pride would soon become his heartbreak.

The dam was constructed in a narrow canyon with steep walls, but the underlying geology had not been thoroughly studied. One abutment rested on ancient landslide material, and the other on fractured schist—both unsuitable for supporting a concrete gravity dam. Over the next two years, cracks and seepage were observed. Inspectors were dispatched, but Mulholland and his team dismissed the concerns.

Just before midnight on March 12, a small leak turned into a torrent. The dam failed catastrophically, releasing 12.4 billion gallons of water. The flood traveled 54 miles to the Pacific Ocean, destroying everything in its path.

The aftermath was both personal and institutional. Mulholland took full responsibility, famously saying, "Don't blame anybody else, you just fasten it on me." Though cleared of criminal negligence, he never designed another dam. The disaster led to new dam safety legislation, including the hiring of geologists in all future public works projects in California.

After St. Francis, flood control was no longer just an engineering task—it was a civic responsibility. The disaster swept away lives, but also gave rise to statewide dam regulations, peer review policies, and a deeper understanding that concrete alone does not hold back catastrophe. And nearly a century later, when water again threatened California's Oroville Dam, the ghost of St. Francis stirred—reminding engineers that every reservoir holds two things: pressure, and memory.

Sources (*and related articles*):

- *Wikipedia.* "St. Francis Dam." https://en.wikipedia.org/wiki/St._Francis_Dam
- *U.S. Geological Survey.* "The St. Francis Dam Disaster." USGS photo page. https://www.usgs.gov/media/images/st-francis-dam-disaster
- *ASDSO.* "St. Francis Dam (California, 1928) – Lessons Learned." https://damfailures.org/case-study/st-francis-dam-california-1928/
- *Rogers, J.D.* "Reassessment of the St. Francis Dam Failure." Missouri S&T (1992). https://web.mst.edu/rogersda/temp/St%20Francis%20Dam%20failure/Reassessment%20St%20Francis%20Dam%20Failure-Rogers-1992.pdf

- *Caragozian, J.S.* "The St. Francis Dam: California's Worst Civil Engineering Disaster." California State Archives (2021). https://www.cschs.org/wp-content/uploads/2021/11/History-Resources-Caragozian-St.-Francis-Dam-10-28-21.pdf
- *Geo Forward.* "St. Francis Dam Disaster Site – Facts & Geology." 2024. https://www.geoforward.com/saint-st-francis-dam-geology-geological-disaster-los-angeles-william-mulholland-history/
- *Wired.* "March 12, 1928: To Die in L.A." (2007). https://www.wired.com/2007/03/march-12-1928-to-die-in-l-a

7.
Tacoma Narrows Bridge Collapse (USA, 1940):
When the Wind Learned to Dance

On the morning of November 7, 1940, the Tacoma Narrows Bridge—nicknamed "Galloping Gertie"—twisted and bucked like a living creature before disintegrating spectacularly into the waters of Puget Sound. The bridge, only four months old, had been hailed as an elegant engineering marvel. Its collapse, captured in ghostly black-and-white footage, would become one of the most haunting and instructive events in civil engineering history.

The failure wasn't due to material flaws or overloading. Instead, it was the invisible handshake between structure and wind—a phenomenon called aeroelastic flutter. The bridge's narrow deck, shallow plate girders, and lack of adequate wind bracing made it prone to torsional oscillation. As steady 40 mph winds swept through the Narrows, the bridge began to undulate rhythmically, then violently. Cables snapped. Roadways twisted. And within hours, the bridge tore itself apart.

Amazingly, there were no human fatalities. The only casualty was a terrified dog, left behind in an abandoned car as the bridge collapsed.

But the collapse shook the engineering world. It led to a paradigm shift in bridge design, particularly in the fields of aerodynamics and structural dynamics. Wind tunnel testing became standard practice, and the behavior of flexible structures under dynamic loads was studied with renewed urgency. The tragedy helped birth modern aeroelastic theory—and ultimately made future suspension bridges safer and more wind-resilient.

"Galloping Gertie" did more than fall—it transformed engineering from intuition to science.

"Galloping Gertie" taught engineers to respect the rhythm of wind—but decades later, pedestrians would give them a new tempo to study. In 2000, the Millennium Bridge in London began to sway—not from nature, but from synchronized human footsteps. The bridge held, but the reminder stood firm: motion always finds a way in. (Note: Millennium Bridge's "wobble" due to pedestrian-induced lateral vibration—though nowhere near a collapse—was still a flutter-inspired caution.)

Sources (*and related articles*):

- *Wikipedia.* "Tacoma Narrows Bridge (1940)." https://en.wikipedia.org/wiki/Tacoma_Narrows_Bridge_(1940)
- *University of Washington Libraries.* "Galloping Gertie: The Tacoma Narrows Bridge collapse." https://www.lib.washington.edu/specialcollections/collections/exhibits/tnb
- *Miller, F. J. & Fuller, R. G. (1963).* "Tacoma Narrows Bridge Collapse." PDF documentary script. https://www.aapt.org/store/upload/tacoma_narrows2.pdf

- *Billah, K. Y. & Scanlan, R. H. (1991).* "Resonance, Tacoma Narrows Bridge Failure…" American Journal of Physics. https://courses.washington.edu/cee517/scanlanTN.pdf
- *Arioli, G. & Gazzola, F. (2011).* "Old and new explanations of the Tacoma Narrows Bridge collapse." International Journal of Bifurcation and Chaos. https://gazzola.faculty.polimi.it/aimeta.pdf
- *PDH STAR (2019).* "Ethical Issues from the Tacoma Narrows Bridge Collapse." https://pdhstar.com/wp-content/uploads/2019/01/ELR-052-Ethics-Case-Study-Tacoma-Narrows-Bridge-FINAL.pdf
- *Journal of Emerging Trends and Novel Research (2024).* "A Comprehensive Review of the Aeroelastic Collapse… " https://rjpn.org/jetnr/papers/JETNR2401005.pdf

Part II:
Postwar Reckonings
(1950–1999)

8.
Vajont Dam Disaster (Italy, 1963):
Echoes from a Silent Dam

On the night of October 9, 1963, an eerie silence preceded catastrophe in the Piave Valley of northern Italy. Just before 11:00 PM, a massive landslide—some 260 million cubic meters of rock—broke free from Monte Toc and thundered into the Vajont reservoir at astonishing speed, displacing the water like a piston in a barrel. A wave nearly 250 meters high surged over the dam's crest, not breaking it but vaulting over it, annihilating everything in its path.

The flood swept down into the towns of Longarone, Pirago, Rivalta, and others, erasing communities in seconds. Nearly 2,000 lives were lost—many bodies were never recovered. Homes, schools, and entire generations vanished beneath the wave.

Remarkably, the Vajont Dam itself remained standing—a monument not of failure, but of misused engineering. The disaster wasn't structural; it was geological—and human. Geologists had warned for years that Monte Toc's slope was unstable, and early signs of creeping and fissures were well documented. But the company behind the project, SADE (Società Adriatica di Elettricità), chose to ignore or downplay concerns. Test results were misinterpreted or buried. The risk was known—it was simply inconvenient.

The tragedy led to widespread public outrage, legal proceedings, and deep reflection on the ethical responsibilities of engineers, geologists, and government regulators. It remains a textbook example in engineering ethics, geotechnical due diligence, and the fatal cost of silencing dissent.

The dam held. But trust, conscience, and an entire valley did not.

In the decades that followed, atrium designs became synonymous with elegance—but also with caution. No major collapse echoed the Hyatt Regency's exact flaw, but every suspended structure since has carried its memory, embedded like a hidden load in the blueprint.

Sources (*and related articles*):

- *Wikipedia*, "Vajont Dam." https://en.wikipedia.org/wiki/Vajont_Dam
- *ASDSO (via damfailures.org)*, "Vajont Dam (Italy, 1963)." https://damfailures.org/case-study/vajont-dam-italy-1963/
- *The 1963 Vajont Landslide: 50th Anniversary*, ResearchGate (2013). https://www.researchgate.net/publication/258160480_The_1963_Vajont_Landslide_50th_Anniversary
- *ISSMGE*, "Lessons from the Vaiont landslide" PDF. https://www.issmge.org/uploads/publications/105/106/ISL2020-4.pdf

- *Springer (2018)*, "New Interpretation of the Vaiont landslide." https://link.springer.com/article/10.1007/s10346-018-0998-9
- *Nature Scientific Reports*, "3D simulation of landslide-water dynamics." https://www.nature.com/articles/s41598-024-77922-5
- *ASCE Library*, "Vajont Disaster: SPH Modeling." https://ascelibrary.org/doi/10.1061/%28ASCE%29HY.1943-7900.0001111
- *University of Warwick*, "1963 Vajont rock slide." https://warwick.ac.uk/fac/sci/eng/research/grouplist/ground/geo/publications/geotechnique_vajoint_pre-print.pdf
- *University of Colorado*, "Vaiont Reservoir Disaster" PDF. https://ceae.colorado.edu/~amadei/CVEN5768/PDF/Vaiont%20Reservoir%20Disaster.pdf

9.
Ronan Point Collapse (UK, 1968):
When Walls Were Only Holding On

On the morning of May 16, 1968, an East London resident lit her stove in the newly opened Ronan Point high-rise. A gas leak ignited a small explosion—minor by traditional standards—but in that moment, the building revealed its terrifying fragility. A single exterior wall panel blew out on the 18th floor, and with it, an entire corner of the building peeled away. Four people died instantly, crushed in their homes as floor after floor pancaked down in a chain reaction.

The Ronan Point tower was built using the "large panel system"—a form of precast, prefabricated concrete construction where wall and floor panels were simply bolted together. It was cost-effective, fast, and part of a wave of post-war public housing initiatives across Britain. But the explosion exposed a fatal lack of structural redundancy. Once one panel failed, there was no robust framework to redistribute the load. The building fell like a house of cards.

The disaster rattled public faith in modernist architecture and prefab tower blocks. Though only a few people died, the symbolism was seismic: if one small gas leak could destroy a tower, how safe were the hundreds of similar buildings across the UK?

The event led to sweeping reforms in building regulations, particularly around progressive collapse resistance, gas safety in high-rises, and the minimum standards for structural interdependence in residential architecture. Ronan Point became a shorthand for what happens when efficiency outruns safety, and when cost-saving trumps resilience.

Decades later, the shadow of Ronan Point stirred again—this time as Grenfell Tower burned through the night—proof that when safety is compromised for speed or cost, history may collapse more than once.

Sources (*and related articles*):

- Wikipedia. "Ronan Point." https://en.wikipedia.org/wiki/Ronan_Point
- *Government Inquiry Report (1968)* [PDF]. https://ia801209.us.archive.org/15/items/op1268013-1001/op1268013-1001.pdf
- *ResearchGate.* "Ronan Point Apartment Tower Collapse and Its Effect on Building Codes." https://www.researchgate.net/publication/245290238_Ronan_Point_Apartment_Tower_Collapse_and_its_Effect_on_Building_Codes
- *PCI Journal (1975).* "Design against progressive collapse." https://www.pci.org/PCI_Docs/Publications/PCI%20Journal/1975/March-1975/Design%20Against%20Progressive%20Collapse.pdf

- *Warwick University*. "Historical review of prescriptive design rules for robustness after the Ronan Point collapse." https://wrap.warwick.ac.uk/id/eprint/117589/1/WRAP-historical-review-prescriptive-design-rules-robustness-collapse-Ronan-Point-Russell-2019.pdf

- *Failure Knowledge Database*. "Chain Reaction Collapse of a High-rise Apartment due to a Gas Explosion." https://www.shippai.org/fkd/en/hfen/HA1000634.pdf

- *Buildingsafetyregister.co.uk*. "Ronan Point at 57: The Collapse That Reshaped Building Regulations." https://buildingsafetyregister.co.uk/ronan-point-at-57-the-collapse-that-reshaped-building-regulations/

10.
West Gate Bridge Collapse (Australia, 1970):
When Compression Meets Complacency

Although this predates the 21st century, the collapse of the West Gate Bridge in Melbourne, Australia, on October 15, 1970, remains a defining example of engineering oversight.

During construction, a span of the bridge buckled due to uneven stresses between sections being joined. Engineers attempted to correct the camber by removing bolts, but the sudden redistribution of load caused the 2,000-ton section to collapse, killing 35 workers.

The event led to changes in bridge construction practices, including improved stress analysis methods and clearer communication between site engineers and design consultants.

The West Gate Bridge fell not just from steel misaligned, but from voices unheard and warnings unheeded. Its twisted span became a landmark not of connection, but of consequence—reminding engineers that design is only as safe as its execution. Decades later, echoes of its failure could still be heard in the Kellogg Bridge collapse (USA, 1970) and Yekaterinburg footbridge collapse (Russia, 2006)—each a painful reminder that temporary stages in construction are never temporary in responsibility.

Sources (*and related information*):

- *Wikipedia.* "West Gate Bridge." https://en.wikipedia.org/wiki/West_Gate_Bridge

- *Create Digital.* "West Gate Bridge Collapse: Engineering Lessons." https://createdigital.org.au/west-gate-bridge-collapse-engineering-lessons/

- *ABC News (Callanan, Tim).* "Disaster still haunts survivors after 50 years." Oct 11, 2020. https://www.abc.net.au/news/2020-10-11/west-gate-bridge-collapse-haunts-survivors-50-years-on/12739324

- *Museums Victoria.* "Remembering the West Gate Bridge Collapse, 15 October 1970." https://collections.museumsvictoria.com.au/articles/17000

- *SheerForce Engineering.* "Why Did the West Gate Bridge Collapse?" Oct 26, 2021. https://sheerforceeng.com/2021/10/26/why-did-the-west-gate-bridge-collapse/

- *Monash University Lens.* "50 years on: lessons learned." Oct 14, 2020. https://lens.monash.edu/@technology/2020/10/14/1381519/remembering-the-west-gate-bridge-collapse-50-years-on-and-what-we-learnt

11.
Flixborough Disaster (UK, 1974):
A Chemical Chain Reaction

On June 1, 1974, a temporary pipe at a chemical plant in Flixborough exploded, killing 28 people and injuring more than 100. The blast devastated the plant and nearby homes, and the fire burned for days.

The disaster stemmed from an untested bypass pipe installed to keep production going after a reactor crack. Engineers failed to analyze the mechanical stress and compatibility of the substitute system. When it failed, a large release of cyclohexane ignited instantly.

The Flixborough explosion catalyzed sweeping changes in industrial safety laws in the UK, including the formation of the Health and Safety Executive (HSE).

Flixborough exploded not just from pressure, but from a decision made without drawings, calculations, or second thought. The blast tore through more than steel—it ripped open the hidden dangers of temporary fixes in permanent systems. Its legacy would echo years later in Seveso (Italy, 1976), Texas City (USA, 2005), and Toulouse (France, 2001)—each a brutal reminder that in chemical engineering, there is no such thing as a harmless shortcut.

Sources (*and related articles*):

- *Wikipedia.* "Flixborough Disaster." https://en.wikipedia.org/wiki/Flixborough_disaster
- *HSE.* "Flixborough (Nypro UK) Explosion 1st June 1974 – Accident Summary." https://www.hse.gov.uk/comah/sragtech/caseflixboroug74.htm
- *IChemE Loss Prevention Bulletin #289.* https://www.icheme.org/media/19550/lpb289_pg24.pdf
- *Inspectioneering.com.* "The Start of Process Safety Management: The Flixborough Disaster." https://inspectioneering.com/blog/2022-02-24/10030/lets-be-frank-the-start-of-process-safety-management-the-flixborough-disaster
- *ARIA.* "Catastrophic explosion of a cyclohexane cloud." https://www.aria.developpement-durable.gouv.fr/wp-content/files_mf/FD_5611_flixborough_1974_ang.pdf
- *ScienceDirect.* "Flixborough: A final footnote." https://www.sciencedirect.com/science/article/abs/pii/S0950423007000800
- *YouTube.* "Flixborough Disaster Nypro 1974." https://www.youtube.com/watch?v=lwCvVW1oe6Y

- *Finch Consulting.* "51 years on from the Flixborough disaster …" https://www.finch-consulting.com/51-years-on-from-the-flixborough-disaster-what-happened-and-what-have-we-learnt/
- *IChemE Incident Summary (2023).* https://www.icheme.org/media/14091/flixborough-incident-summary-01-jun-74.pdf
- *NoonPi blog.* "The Flixborough Works of Nypro Explosion and Fire." https://noonpi.com/the-flixborough-works-of-nypro-explosion-and-fire/

12.
Banqiao Dam Failure (China, 1975):
The Hidden Hydrological Horror

Although this occurred in the 20th century, the Banqiao Dam collapse is essential in any list of engineering failures. On August 8, 1975, after days of unprecedented rainfall from Typhoon Nina, the Banqiao Dam in China's Henan Province failed catastrophically.

Designed to withstand a "1-in-1,000-year flood," the dam was overwhelmed due to flawed hydrological models and political pressure to store excessive water. The collapse triggered a domino effect, breaching 61 other dams and resulting in over 170,000 deaths, many from subsequent disease and famine.

Kept secret for decades, Banqiao remains one of the deadliest man-made disasters in history and a sobering lesson in the intersection of politics and engineering.

Banqiao did not fail alone. Its collapse triggered a cascade—of water, of dams, of silence. Built to tame nature, it was undone by politics, pressure, and rainfall no one dared plan for. Later disasters—from Machchu II (India, 1979) to Situ Gintung (Indonesia, 2009)—would reveal a similar truth: that no wall is higher than the warnings it ignores, and no reservoir ever holds back a storm indefinitely.

Sources (*and related articles*):

- *Wikipedia.* "1975 Banqiao Dam failure." https://en.wikipedia.org/wiki/1975_Banqiao_Dam_failure

- *DamFailures.org (ASDSO).* Banqiao Dam Case Study. https://damfailures.org/case-study/banqiao-dam-china-1975/

- *CBDB*. "FTA Analysis of Banqiao Dam Failure." *Brazil Seminar 2023*. https://cbdb-web-uploads.s3.amazonaws.com/system/uploads/ckeditor/attachments/1265/078_-_FTA_ANALYISIS_OF_BANQUIAO_DAM_FAILURE.pdf

- *Encyclopaedia Britannica*, "Typhoon Nina–Banqiao dam failure." https://www.britannica.com/event/Typhoon-Nina-Banqiao-dam-failure

- *Bowdoin College*. "Banqiao Dam Failure: Series of Events." https://courses.bowdoin.edu/history-2203-fall-2020-whausman/narrative-of-the-event/

- *SHEQ Network*. "The Banqiao Dam Disaster and Contractor Management Lessons." https://sheqnetwork.com/2024/11/26/the-banqiao-dam-disaster-and-the-importance-of-a-robust-contractor-management-system-lessons-for-modern-engineering-projects/

- *ASCE (2008).* "Lessons from catastrophic dam failures in August 1975 in Zhumadian, China." https://ascelibrary.org/doi/10.1061/40971%28310%2920
- *AMS Journals.* "Typhoon Nina and the August 1975 Flood." https://journals.ametsoc.org/view/journals/hydr/18/2/jhm-d-16-0152_1.xml

13.
Three Mile Island Partial Meltdown (USA, 1979):
A Near Nuclear Nightmare

On March 28, 1979, the Three Mile Island nuclear power plant in Pennsylvania, USA, experienced a partial meltdown of its reactor core. A stuck valve combined with misleading control room indicators caused operators to misjudge the problem, worsening the situation.

Although no direct deaths occurred, the incident exposed serious design flaws, including over-reliance on automation and a poor human-machine interface. Inadequate training, complex systems, and poor alarm management were also blamed.

This incident drastically slowed the growth of nuclear energy in the U.S., led to the Nuclear Regulatory Commission's reforms, and contributed to stronger plant safety standards worldwide.

Three Mile Island didn't explode—but it shook the foundations of nuclear confidence. In its quiet core lay a loud message: that even without fire or fallout, confusion and complacency can bring a system to its knees. Years later, Chernobyl (USSR, 1986) and Fukushima (Japan, 2011) would turn that message into a warning shouted across continents—and prove that in nuclear engineering, near misses are never near enough.

Sources (*and related articles*):

- *Wikipedia.* "Three Mile Island accident." https://en.wikipedia.org/wiki/Three_Mile_Island_accident
- *U.S. Nuclear Regulatory Commission*, "Backgrounder on the Three Mile Island Accident." https://www.nrc.gov/reading-rm/doc-collections/fact-sheets/3mile-isle.html
- *World Nuclear Association*, "Three Mile Island Accident." https://www.world-nuclear.org/information-library/safety-and-security/safety-of-plants/three-mile-island-accident.aspx
- *Environmental Justice / EJHistory*, "Three Mile Island Media." https://ejhistory.com/three-mile-island-media/

14.
Hyatt Regency Walkway Collapse (USA, 1981):
A Deadly Design Shortcut

On July 17, 1981, two suspended walkways collapsed in the atrium of the Hyatt Regency hotel in Kansas City, Missouri, during a tea dance attended by hundreds. The collapse killed 114 people and injured more than 200, making it one of the deadliest structural failures in U.S. history.

The disaster was caused by a critical engineering design change: instead of a single-rod support, the contractor used a double-rod system that doubled the load on a single connection point. The change was not properly analyzed or approved.

This tragedy reshaped engineering ethics, led to stricter peer review practices, and is now taught as a cautionary tale in structural engineering curricula worldwide.

In the decades that followed, atrium designs became synonymous with elegance—but also with caution. No major collapse echoed the Hyatt Regency's exact flaw, but every suspended structure since has carried its memory, embedded like a hidden load in the blueprint.

Sources (*and related articles*):

- *Wikipedia.* "Hyatt Regency walkway collapse." https://en.wikipedia.org/wiki/Hyatt_Regency_walkway_collapse

- *ASCE Civil Engineering Source (2007).* "Design Alternates." https://ascelibrary.org/doi/10.1061/(ASCE)0733-9445(1983)109:5(1226)
- *UNT Library.* "Hyatt Regency Walkways Collapse." https://digital.library.unt.edu/ark:/67531/metadc38324/m2/1/high_res_d/metadc38324.pdf
- *Texas A&M Ethics (2017).* "Hyatt Regency Walkway Collapse." https://ethics.tamu.edu/wp-content/uploads/sites/7/2017/04/HyattRegency.pdf
- *YouTube.* "Hyatt Regency Walkway Collapse: Ethics Case Study No. 4." https://www.youtube.com/watch?v=98l3Us8IU-I

15.
Bhopal Gas Tragedy (India, 1984):
The Night of Poisoned Air

Shortly after midnight on December 3, 1984, a deadly cloud of methyl isocyanate (MIC) gas leaked from the Union Carbide pesticide plant in Bhopal, India. Over 3,000 people died within hours; thousands more died in the following weeks, and hundreds of thousands suffered long-term health effects.

The disaster was due to a cascade of engineering and organizational failures—faulty safety systems, under-maintained equipment, lack of scrubbers, disabled alarms, and insufficient containment. Engineers had warned of hazards months earlier, but cost-cutting and managerial neglect overrode those concerns.

The Bhopal disaster remains the world's worst industrial accident and a turning point in chemical plant safety and international corporate accountability.

The gases have long dispersed, but Bhopal's silence still speaks. It warns that when profit outweighs accountability, and safety becomes a footnote, the cost is measured not in dollars or rupees—but in generations.

Bhopal remains a wound, not a memory. Its lessons were carved in law, protest, and policy—yet echoes still surface. In 2020, a toxic gas leak in Visakhapatnam revived old fears: that safeguards are only as strong as the urgency behind them, and that the world forgets more easily than it learns.

Sources (*and related articles*):

- *Wikipedia* – "Bhopal disaster." https://en.wikipedia.org/wiki/Bhopal_disaster
- *Wikipedia* – "International Medical Commission on Bhopal." https://en.wikipedia.org/wiki/International_Medical_Commission_on_Bhopal
- *Wikipedia* – "Rajkumar Keswani." https://en.wikipedia.org/wiki/Rajkumar_Keswani
- *Wikipedia* – "Sambhavna Trust Clinic." https://en.wikipedia.org/wiki/Sambhavna_Trust_Clinic
- *Wikipedia* – "Five Past Midnight in Bhopal." https://en.wikipedia.org/wiki/Five_past_Midnight_in_Bhopal
- *The Bhopal Action Group*. "The Disaster." https://www.bhopal.org/the-disaster/
- *Prasad, S. et al.* "The Bhopal disaster and its aftermath: a review." *Indian J Med Res.* 2003. https://www.ncbi.nlm.nih.gov/pmc/articles/PMC1142333/

16.
Hotel New World Collapse (Singapore, 1986):
The weight we forgot to carry

On the morning of 15 March 1986, the six-story Lian Yak Building—popularly known as Hotel New World—stood quietly at the intersection of Serangoon and Owen Roads in Singapore's Little India. Completed in 1971, it housed a bank, a nightclub, and 67 hotel rooms above. At 11:25 a.m., with no warning and without seismic activity, the entire structure pancaked in under a minute. The collapse obliterated every column and wall, burying occupants and passers-by beneath concrete and steel. Rescue teams, including the Civil Defence Force, Singapore Armed Forces, and international experts, worked tirelessly for days. Of the estimated 300 caught under rubble, 33 people died, and 17 were miraculously rescued.

The subsequent inquiry revealed a chilling oversight: the structural engineer omitted the dead load—the weight of the building itself—from calculations, only accounting for live load. Over time, three columns had developed microcracks, heightened by added rooftop equipment and surfaced headers. Finally, the columns failed, initiating progressive collapse throughout the building.

The disaster sent ripples through Singapore's construction regulations. In response, the government mandated independent accredited structural checks and periodic safety surveys for older buildings—transformational changes meant to ensure such a failure would never recur

Though decades have passed, the Hotel New World disaster remains an enduring lesson in structural responsibility. It shows how a calculation left unmade can topple a city's sense of safety, and why every formula matters when gravity has the final say—no matter where you build in the world. Similar missteps have claimed lives elsewhere: in Nice, France (1973), an apartment block collapsed due to concrete fatigue and under-designed support columns; in Campinas, Brazil (2007), a church roof caved in during mass due to design flaws in its steel truss system; and in Madrid, Spain (2006), a parking garage under construction failed catastrophically when its slab design could not sustain the intended loads.

Sources (*and related articles*):

- *Wikipedia.* "Collapse of Hotel New World." https://en.wikipedia.org/wiki/Collapse_of_Hotel_New_World
- *Wikipedia.* "Progressive collapse." https://en.wikipedia.org/wiki/Progressive_collapse
- *National Heritage Board.* "Collapse of Hotel New World." https://www.roots.gov.sg/stories-landing/stories/collapse-of-hotel-new-world/story
- *The Straits Times* (Article archive, Singapore Press Holdings – login required) https://eresources.nlb.gov.sg/newspapers/Digitised/Article/straitstimes19860317-1.2.2

- *Roots Magazine* (PDF archive via NHB).
  "Muse Issue 2" includes retrospective on the Hotel New World collapse—downloadable PDF from archives.nhb.gov.sg. Page 16.
  https://www.roots.gov.sg/-/media/Roots/Images/resources/muse/issues/030-muse-v8-issue2.pdf

17. Chernobyl Nuclear Disaster (USSR, 1986):

Catastrophe at Reactor No. 4

On April 26, 1986, during a late-night safety test at the Chernobyl Nuclear Power Plant in Ukraine (then part of the Soviet Union), Reactor No. 4 exploded. The blast released massive amounts of radioactive material across Europe.

Chernobyl was caused by a flawed reactor design, a poorly executed test, and a culture of secrecy. The RBMK reactor lacked proper containment structures, and operators disabled safety systems during the test. The explosion killed two people instantly and exposed thousands to radiation. Long-term effects include increased cancer rates and environmental devastation.

The disaster marked a pivotal moment in nuclear energy policy and engineering safety, triggering sweeping reforms and global attention to reactor design and disaster preparedness.

Chernobyl's reactor core burned for days, but its legacy will glow for centuries. Long after the sarcophagus was sealed, its warnings still leak—into engineering handbooks, energy policies, and every risk model that dares to ask, "What if we push just a little further?"

Sources (*and related articles*):

- *Wikipedia.* "Chernobyl disaster." https://en.wikipedia.org/wiki/Chernobyl_disaster

- *Wikipedia.* "Effects of the Chernobyl disaster." https://en.wikipedia.org/wiki/Effects_of_the_Chernobyl_disaster
- *Wikipedia.* "Chernobyl exclusion zone." https://en.wikipedia.org/wiki/Chernobyl_exclusion_zone
- *World Nuclear Association.* "Chernobyl Accident." https://www.world-nuclear.org/information-library/safety-and-security/safety-of-plants/chernobyl-accident.aspx
- *U.S. NRC.* "Backgrounder on the Chernobyl Accident." https://www.nrc.gov/reading-rm/doc-collections/fact-sheets/chernobyl-bg.html
- *Britannica.* "Chernobyl Disaster." https://www.britannica.com/event/Chernobyl-disaster
- *IAEA FAQs.* "What caused the Chernobyl accident?" https://www.iaea.org/newscenter/focus/chernobyl/faqs
- *UNSCEAR.* "The Chernobyl Accident." https://www.unscear.org/unscear/en/chernobyl.html

18.
Seongsu Bridge Collapse (South Korea, 1994):
Metal Fatigue and Missed Warnings

On October 21, 1994, during morning rush hour in Seoul, a central section of the Seongsu Bridge collapsed into the Han River, killing 32 people. The collapse shocked the nation and spurred an overhaul of infrastructure inspection practices.

The bridge failed due to metal fatigue and poor welding in a load-bearing truss. Investigations revealed inadequate maintenance, rushed construction, and failure to act on previous safety warnings.

This incident reshaped South Korea's civil engineering protocols and prompted investment in more robust bridge inspection regimes.

The Seongsu Bridge didn't fall from wind, flood, or earthquake—it fell from within. Rusted joints, weak welds, and a system too hurried to inspect turned a routine commute into catastrophe. In its fallen span lay a country's reckoning: that progress built too fast, and maintained too little, becomes its own risk. Years later, when the I-35W Bridge collapsed in Minneapolis (USA) and the Morbi suspension bridge failed in Gujarat (India), the message resurfaced—quiet infrastructure is not safe infrastructure, unless vigilance is louder than urgency.

Sources (*and related articles*):

- *Wikipedia.* "Seongsu Bridge disaster." https://en.wikipedia.org/wiki/Seongsu_Bridge_disaster

- *Scribd.* "Forensic Structural Engineering – Case: Seongsu Bridge." https://www.scribd.com/document/472473264/0-1

- *LinkedIn.* "Lessons from the Seongsu Bridge Collapse – A Civil Engineer's Perspective." https://www.linkedin.com/pulse/lessons-from-seongsu-bridge-collapse-civil-engineers-perspective-8ctqf

- *Failure Knowledge Database.* "Collapse of the Seongsu Bridge." https://www.shippai.org/fkd/en/cfen/CD1000144.html

- *The Korea Herald.* "Bridge collapse lays bare dark side of Han River infrastructure." https://www.koreaherald.com/article/3175790

19. Sampoong Department Store Collapse (South Korea, 1995): *A Price Too High for Extra Floors*

On a scorching summer afternoon, June 29, 1995, Seoul's Sampoong Department Store—an iconic pastel-pink landmark—hummingly served thousands of shoppers in its five-story grandeur. Behind its modern façade lurked a deadly secret. Though initially conceived as a four-story office block, the building had been reconfigured mid-construction into a department store. Its owner, eager to maximize profit, pushed for a fifth floor and escalators, even in the face of severe warnings from initial contractors.

That afternoon, subtle sounds of distress echoed through the shopping halls: the ceiling of the fifth floor sagged, faint cracks snaked across the concrete, and heavy vibrations from roof-mounted air-conditioning units rattled the structure. But limited inspections ended with a token evacuation of the top floor—and grimly, management chose profits over prudence, keeping the rest of the crowd inside.

At 17:57, the inevitable occurred. A surge of failure cascaded through the building's weakened skeleton. In mere 20 seconds, structural columns gave way under the strain of extra weight, missing reinforcements, and altered load paths. The shopping center pancaked inwards—five stories collapsing floor onto floor—burying an estimated 1,500 people beneath the rubble.

In the deafening silence that followed, cranes and rescue teams combed through twisted steel and concrete dust. Miraculously, 17 survivors were pulled alive from the wreckage, but 502 lives had been lost, and 937 injured—among them children, foreign tourists, and store employees. The scale of devastation was unprecedented, marking the deadliest non-terrorist building collapse of the 20th century.

What emerged from the investigation was a catalogue of greed-driven miscalculations: columns were reduced from the designed 80 cm to just 60 cm, while steel reinforcements were halved from 16 bars to 8. Gutting support walls for escalators and dragging 45-tonne AC units across the roof—rather than hoisting them properly—induced fatal stresses in already compromised columns.

Critics argue this tragedy wasn't just technical—it was moral. Financial ambitions drowned structural integrity, and officials who rubber-stamped permits were later convicted of bribery and negligence. Owner Lee Joon and his son were jailed; Seoul responded with sweeping reforms, mandating independent structural audits and bolstering construction oversight.

Though six years have passed since another catastrophe—Rana Plaza—Sampoong remains a stirring reminder of how ambition can outpace engineering, and why effective oversight must stand guard over every beam and bolt. Across the world, when cost trumps caution, structures don't collapse—they betray us.

Sources (*and related articles*):

- *Wikipedia.* "Sampoong Department Store collapse." https://en.wikipedia.org/wiki/Sampoong_Department_Store_collapse

- *Encyclopaedia Britannica.* "Sampoong Department Store collapse." https://www.britannica.com/event/Sampoong-Department-Store-collapse
- *The Guardian.* "Learning from Seoul's Sampoong Department Store disaster." https://www.theguardian.com/cities/2015/may/27/seoul-sampoong-department-store-disaster-history-cities-50-buildings
- *PubMed (Elsevier).* "Inspection of collapse cause of Sampoong Department Store." https://pubmed.ncbi.nlm.nih.gov/22079499/
- *Penn State Engineering Wiki.* "Failures – Sampoong Department Store." https://www.engr.psu.edu/ae/thesis/failures/MKP/failures/failures.wikispaces.com/Sampoong_Department_Store.html

20.
İzmit Earthquake Infrastructure Failures (Turkey, 1999):
When Structures Were Not Built to Last

In the early morning of August 17, 1999, a 7.6-magnitude earthquake struck the Marmara region of northwestern Turkey, killing over 17,000 people and displacing more than 250,000. While the tremor itself was powerful, much of the devastation was due to widespread structural failures—buildings, bridges, and industrial sites collapsed en masse, many of them less than a decade old.

The city of İzmit and surrounding towns were riddled with reinforced concrete structures that had not adhered to seismic design codes. Investigations revealed that substandard materials, poorly enforced regulations, and corruption had allowed contractors to cut corners. Many structures lacked proper reinforcement or used inadequate concrete mixes.

Among the most tragic failures were entire apartment complexes that collapsed pancake-style, burying families alive. Industrial damage included the Tupras oil refinery fire, which burned for days and caused ecological damage.

The İzmit disaster led to intense scrutiny of Turkey's construction industry, overhauls in building codes, and the formation of emergency response and urban resilience frameworks. Yet it remains a sobering reminder of what happens when engineering standards are ignored in earthquake-prone zones.

The ground moved for less than a minute, yet what collapsed had been weakening for years. Substandard concrete, missing reinforcements, and overlooked inspections turned natural fault lines into zones of man-made disaster. İzmit was not just a tectonic tragedy; it became a structural confession. And as similar devastation unfolded in Bam (Iran, 2003), Port-au-Prince (Haiti, 2010), and Nepal (2015), the world was reminded once more: while nature may strike first, it is often human error that brings the deadliest consequences.

Sources (*and related articles*):

- *Wikipedia.* "1999 İzmit earthquake." https://en.wikipedia.org/wiki/1999_%C4%B0zmit_earthquake
- *EERI* Special Earthquake Report, October 1999. https://www.eeri.org/lfe/pdf/Turkey_Kocaeli_Insert_Oct99.pdf
- *Bird et al.* (2002). "Site effects and building damage in Adapazarı." BSSA. https://www.sciencedirect.com/science/article/abs/pii/S0141029699001066
- *Bakir & Sucuoglu* (2004). "Izmit earthquake damage estimation." BEER. https://link.springer.com/article/10.1007/s10518-004-3804-0

Part III:
21st Century Failures
(2000–Present)

21.
Basmanny Market Roof Collapse (Russia, 2006):
Snow Load Miscalculated

On February 23, 2006, in the early hours of a frigid Moscow morning, the roof of the Basmanny Market caved in without warning. Beneath the ruins lay 66 dead and over 100 injured—shoppers, vendors, and workers trapped under twisted steel and concrete. Once considered a showcase of Soviet-era engineering, the building's domed steel frame, erected in the 1970s, had long since fallen into neglect.

Investigators found that years of corrosion and fatigue had eaten away at key structural connections. The welds were brittle, inspections sporadic, and the roof's curved design—visually elegant—offered poor resistance to uneven snow accumulation. Compounding the problem was a critical underestimation of dynamic snow loads and little redundancy in the event of progressive failure.

The tragedy led to a wave of safety reforms in Russia's aging public infrastructure, particularly targeting large-span roofs and market halls built under Soviet-era standards. Several similar structures were shut down or demolished, and winter-load protocols were revised.

Basmanny was not brought down by snow alone, but by the quiet erosion of oversight. Deferred inspections, overconfidence in aging designs, and the natural weight of winter turned a shelter into a tomb. Later collapses—

from Katowice (Poland, 2006) to Bad Reichenhall (Germany, also 2006)—would repeat the same lesson: that snow does not forgive, and structures meant to welcome must be designed to endure.

Sources (*and related articles*):

- *Wikipedia.* "Basmanny market roof collapse." https://en.wikipedia.org/wiki/Basmanny_market_roof_collapse
- *ResearchGate.* "Applicability of PSInSAR for building hazard identification … Moscow Basmanny market collapse." https://www.researchgate.net/publication/287094926_Applicability_of_psinsar_for_building_hazard_identification_Study_of_the_29_January_2006_Katowice_exhibition_hall_collapse_and_the_24_February_2006_Moscow_basmanny_market_collapse
- *Tampa Bay Times.* "Moscow market roof collapse kills 56." https://www.tampabay.com/archive/2006/02/24/moscow-roof-collapse-kills-56/
- *WorkingWithCrowds.com.* "Moscow market roof collapse – technical operation findings." https://www.workingwithcrowds.com/moscow-market-roof-collapse-14-02-2004/

22. The Big Dig Tunnel Ceiling Collapse (USA, 2006):

Anchored in Assumption

Boston's Big Dig was envisioned as a triumph of urban engineering—a $14 billion megaproject that buried congested highways beneath the city and stitched together neighborhoods long divided by concrete. But on July 10, 2006, that vision quite literally fell apart. A massive concrete ceiling panel in the Interstate 90 connector tunnel came crashing down onto a moving vehicle, killing 38-year-old Milena Del Valle and narrowly sparing her husband. What should have been a routine evening commute became a fatal consequence of invisible design failure.

The collapse was traced not to the concrete panels themselves, but to the adhesive anchors—epoxy bolts—used to suspend them from the tunnel roof. These bolts, critical to the panel suspension system, were installed in tension, even though they were never rated for long-term, sustained overhead loading. Test data and warnings from the epoxy manufacturer had made this limitation clear, yet engineers, contractors, and project overseers either dismissed or overlooked them. The wrong type of epoxy was used in several locations, and poor installation techniques and inadequate quality control only magnified the risks. The failure wasn't a sudden catastrophe—it was a slow, silent compromise embedded into the tunnel's ceiling from the moment the bolts were installed.

The incident triggered federal investigations, congressional hearings, and criminal indictments, including charges of involuntary manslaughter against the epoxy manufacturer. More broadly, it forced a nationwide reassessment of fastener specifications, tunnel safety standards, and inspection protocols in high-profile infrastructure projects. For a project hailed as a symbol of modern engineering, the collapse exposed how even in the most expensive and technically sophisticated ventures, shortcutting specifications in the fine print can still kill.

The Big Dig ceiling collapse was not a failure of concrete or steel—but of assumptions, oversight, and ignored expertise. Its echoes could be heard in the Minnesota I-35W Bridge collapse (2007), where undersized gusset plates failed silently for years, and in the Gleinalm Tunnel incident (Austria, 2018), where ceiling anchors were found dangerously corroded. Each disaster served as a grim reminder: when safety hangs by a thread—or a bolt—it's not the load that fails first, but the logic behind how it was secured.

Sources (*and related articles*):

- *Wikipedia.* "Big Dig ceiling collapse." https://en.wikipedia.org/wiki/Big_Dig_ceiling_collapse
- *National Transportation Safety Board.* "Ceiling Collapse in the Interstate 90 Connector Tunnel, Boston, Massachusetts, July 10, 2006." NTSB/HAR-07/02. https://www.ntsb.gov/investigations/AccidentReports/Reports/HAR0702.pdf

- *NASA Safety Message.* "Tunnel of Terror – The Big Dig Ceiling Tile Collapse." https://sma.nasa.gov/docs/default-source/safety-messages/safetymessage-2008-06-01-thebigdigceilingtilecollapse.pdf?sfvrsn=e3a91ef8_4
- *Design News.* "Epoxy Creep Main Factor in Big Dig Ceiling Panel Collapse." https://www.designnews.com/assembly/-epoxy-creep-main-factor-in-big-dig-ceiling-panel-collapse
- *Institution of Structural Engineers (IStructE).* "Interstate 90 Connector Tunnel ceiling collapse" PDF. https://www.istructe.org/sitefiles/handlers/downloadfile.ashx?productid=1469

23.
São Paulo Metro Tunnel Collapse (Brazil, 2007):
Collapse Beneath the City

A quiet afternoon in São Paulo turned into chaos on January 12, 2007, when the ground suddenly gave way at the construction site of the city's ambitious Line 4 metro project. What began as a minor surface crack quickly expanded into a gaping crater, large enough to swallow buses, cars, and entire sections of adjacent buildings. In the aftermath, seven people were confirmed dead, and several more were injured. For residents, commuters, and engineers alike, the event was both shocking and deeply unsettling.

At the heart of the disaster was a lethal mix of engineering misjudgments and geotechnical challenges. The construction site sat atop unstable, water-saturated soil that demanded careful excavation, continuous monitoring, and thorough reinforcement—none of which were sufficiently implemented. Waterproofing measures were poorly executed, and the inflow of groundwater had been underestimated by the design team. As water infiltrated the soil and weakened the tunnel's lining, the roof collapsed with devastating force. The Brazilian construction giant Odebrecht, overseeing the project, came under intense public and legal scrutiny for its role in the failure.

Beyond the tragic loss of life, the collapse brought São Paulo's metro expansion to a grinding halt. Line 4, which was intended to ease congestion in one of the world's most populous cities, faced years of delay. The disaster sparked national conversations about the risks of deep urban tunneling, especially in densely populated areas. In response, Brazil introduced stricter regulatory frameworks for underground construction projects, emphasizing soil analysis, real-time monitoring, and accountability in engineering contracts. The hole left in the city's infrastructure was more than physical—it exposed serious gaps in oversight, planning, and the human cost of pushing forward without proper precautions.

The São Paulo tunnel collapse was not an isolated tragedy—it was part of a deeper pattern where excavation meets overconfidence. In Cologne (Germany, 2009), the collapse of a metro tunnel under an archive building killed two and destroyed irreplaceable cultural records. In Shanghai (2003), soil instability during subway construction triggered a chain reaction that toppled buildings. And in Fortaleza (Brazil, 2013), another metro excavation triggered landslides and sinkholes in residential zones. These failures—spanning continents—speak the same truth: when cities dig beneath their own foundations, the first thing that must be reinforced is caution.

Sources (*and related articles*):

- *Wikipedia.* "Pinheiros (São Paulo Metro)." https://en.wikipedia.org/wiki/Pinheiros_%28S%C3%A3o_Paulo_Metro%29
- *Tunnels & Tunnelling.* "A unique metro accident in Brazil." https://www.tunnelsandtunnelling.com/analysis/a-unique-metro-accident-in-brazil/

- *Tunnels & Tunnelling News*. "São Paulo metro shaft collapse." https://www.tunnelsandtunnelling.com/news/so-paulo-metro-shaft-collapse/
- *ResearchGate*. "São Paulo Collapse Report – Lessons from Brazil: Pinheiros examined." https://www.researchgate.net/publication/273320074_Sao_Paulo_Collapse_Report_-_Lessons_from_Brazil_Pinheiros_examined
- *IJME Conference PDF*. "Execution mistakes or failures in NATM tunnel construction." https://www.ijme.us/cd_08/PDF/231%20IT%20302.pdf
- *TRID / New Civil Engineer International*. "NATM used despite failure history." https://trid.trb.org/View/820492

24.
Minneapolis I-35W Bridge Collapse (USA, 2007):
Design Oversight Beneath the Surface

On August 1, 2007, during the evening rush hour in Minneapolis, the I-35W Mississippi River bridge—a major eight-lane artery connecting downtown—suddenly gave way. More than 100 meters of deck collapsed into the river, sending cars, trucks, and construction equipment tumbling down. The collapse killed 13 people, injured 145 others, and shocked a nation that took such everyday infrastructure for granted.

Investigators quickly determined that the cause wasn't environmental, nor a result of fatigue cracking, but a fatal design flaw dating back to the bridge's original construction in the 1960s. The gusset plates—the steel connectors joining beams and trusses—had been undersized from the start. Rated at only half the necessary thickness, they had no margin for error. Over the decades, additional concrete overlays, barriers, lighting, and other retrofits incrementally increased the load beyond what those plates were ever meant to handle. The collapse was not the result of one mistake, but of cumulative oversight, where layered improvements masked foundational weaknesses.

The disaster became a watershed moment in American civil engineering. The Federal Highway Administration launched a nationwide review of bridge gusset designs, and thousands of similar truss bridges were reinspected. It also prompted bipartisan action to increase infrastructure spending, accelerate retrofit programs, and prioritize structural redundancy in future designs. For many Americans, it was the first time bridges were no longer just background—they became visible, vulnerable, and urgent.

The I-35W collapse exposed how even steel can buckle under a paper-thin assumption. Its echoes have surfaced in the Cypress Street Viaduct collapse (California, 1989), where brittle columns shattered during an earthquake, and in Genoa's Morandi Bridge (Italy, 2018), where corroded stays gave way despite decades of warnings. In each case, the flaw was visible in hindsight—buried in blueprints, maintenance logs, or overlooked inspections. When bridges fall, it's rarely for lack of strength—but for lack of listening to the signs beneath the surface.

Sources (*and related articles*):

- *Wikipedia.* "I-35W Mississippi River bridge." https://en.wikipedia.org/wiki/I-35W_Mississippi_River_bridge
- *National Transportation Safety Board.* "Collapse of I-35W Highway Bridge." Report NTSB/HAR-08/03 (PDF). https://www.ntsb.gov/investigations/AccidentReports/Reports/HAR0803.pdf
- *MNopedia (Minnesota Historical Society).* "I-35W Bridge Collapse." https://www3.mnhs.org/mnopedia/search/index/event/i-35w-bridge-collapse

- *Popov, A., et al.* "Numerical investigation of collapse of the Minnesota I-35W bridge." ScienceDirect. https://www.sciencedirect.com/science/article/abs/pii/S0141029613005762
- *EngineeringCivil.com*. "Minneapolis I-35W bridge collapse: engineering evaluation." https://www.engineeringcivil.com/minneapolis-i-35w-bridge-collapse-engineering-evaluations-and-finite-element-analysis.html
- *FireRescue1*. "NTSB: Design errors factor in Minn. bridge collapse." https://www.firerescue1.com/mci-mass-casualty-incidents/articles/ntsb-design-errors-factor-in-minn-bridge-collapse-ieQNZiEbfmZeQ6hl/
- *Roads & Bridges*. "NTSB releases report on I-35W bridge collapse." https://www.roadsbridges.com/home/news/10584246/ntsb-releases-report-on-i-35w-bridge-collapse
- *Brady Heywood Consulting*. "Prediction of Bridge Behavior: Case Study." https://www.bradyheywood.com.au/wp-content/uploads/2020/04/52fa7d_928543ad77f24016a6d39aca72929b66.pdf
- *PBS Blueprint America*. "The Dig ~ I-35W Bridge Collapse: NTSB stresses connection between design, weight." https://www.pbs.org/wnet/blueprintamerica/blogs/the-dig-i-35w-bridge-collapse-ntsb-stresses-connection-between-bad-design-and-weight/239/
- *ResearchGate*. "A Case Study of the Minnesota I-35W Bridge Collapse." https://www.researchgate.net/publication/271367759_Prediction_of_Bridge_Behavior_Through_Failure_A_Case_Study_Of_The_Minnesota_I-35W_Bridge_Collapse

25.
Collapse of Cần Thơ Bridge (Vietnam, 2007):
Where Foundations Whispered Failure

On the morning of September 26, 2007, a large section of the partially built Cần Thơ Bridge approach ramp collapsed in southern Vietnam. A 90-metre span, supported by temporary pillars, sheared off and plunged over thirty metres onto a small island beneath it. Approximately 250 workers and engineers were on or under the structure. Between 54 and 60 workers died and more than 80 sustained serious injuries—some buried beneath crushing concrete and twisted steel cables.

At the time, Cần Thơ Bridge was under construction by Japanese firms Taisei, Kajima, and Nippon Steel, funded by an official development loan. The collapse occurred shortly after rain-soaked soil undermined one of the temporary pile foundations. Investigators found that a temporary pillar had slipped due to uneven settlement of soft sand, causing structural supports to fail in a domino pattern. Design miscalculations, inadequate soil surveys, and cost-cutting in shoring protocols compounded the hazard, leading to catastrophic failure of the approach section.

The Vietnamese government ordered an inquiry and temporarily banned the contracting firms from further projects domestically. The incident delayed completion by several years and prompted stricter enforcement of foundation testing, temporary support standards, and worker safety oversight in high-risk infrastructure sites.

This tragedy underscores a recurring lesson in modern engineering: whether in bridge spans or high-rise towers, settlement underestimations and compromised temporary structures can lead to instant collapse. Similar patterns emerged in São Paulo's metro excavation failures and the Champlain Tower tragedy—when temporary or partial design assumptions fail, the results are irreversible consequences.

Sources (*and related articles*):

- *Wikipedia*. "Collapse of Cần Thơ Bridge." https://en.wikipedia.org/wiki/Collapse_of_C%E1%BA%A7n_Th%C6%A1_Bridge
- *Reuters*. "Bridge collapse kills up to 60 workers in Vietnam." https://www.reuters.com/article/world/up-to-60-killed-in-vietnam-s-worst-bridge-accident-idUSHER626082/
- *The Guardian*. "52 killed in Vietnam bridge collapse." https://www.theguardian.com/world/2007/sep/26/1
- *Los Angeles Times (AP)*. "More than 40 workers dead in Vietnam bridge collapse." https://www.latimes.com/archives/la-xpm-2007-sep-27-fg-bridge27-story.html
- *KHL Group*. "Inquiry reports on Vietnam's Cần Thơ Bridge collapse." https://www.khl.com/news/inquiry-reports-on-vietnam%E2%80%99s-can-tho-bridge-collapse/1026047.article

26. Estádio da Fonte Nova Collapse (Brazil, 2007): *Cracks in the Celebration*

The air in Salvador, Bahia, was electric on November 25, 2007, as fans packed into Estádio da Fonte Nova to celebrate their team's promotion to Brazil's top football division. Joy turned to horror in an instant when a section of the upper terrace gave way, sending dozens of spectators plunging nearly 20 meters to the concrete below. Seven lives were lost, and over 40 others were injured—many of them young fans who had come to witness what should have been a triumphant moment.

Built in 1951, the stadium had long shown signs of aging. Years of exposure to weather and vibrations from heavy crowds had taken a toll on its reinforced concrete structure. Engineers had flagged sections of the upper stands as structurally unsound well before the incident, citing visible cracks and material degradation. However, safety warnings were repeatedly ignored, and no serious efforts were made to reinforce or cordon off vulnerable areas. The tragedy was not a surprise to those familiar with the stadium's condition—it was a disaster waiting to happen.

In the aftermath, the stadium was immediately closed and soon demolished. The collapse sparked a national reckoning about the state of Brazil's aging sports infrastructure, especially in light of the country's preparation for the 2014 FIFA World Cup. A new, modern arena was eventually constructed in its place, but the scars of that day lingered—reminding Brazil and the world that even moments of joy can turn tragic when celebration stands atop unchecked decay. In sports, as in engineering, the foundations must be strong—because lives depend on them.

The collapse at Fonte Nova wasn't just the end of a stadium—it was a collapse of trust in spaces meant to unite and uplift. Similar tragedies have darkened arenas elsewhere: in Abidjan (Ivory Coast, 2009), stampeding fans fell through weak fencing at a World Cup qualifier; in Monterrey (Mexico, 2010), a concert stage roof failed due to poor anchoring and high winds; and in Calcutta (India, 1997), part of the Eden Gardens stadium gave way under overcrowding and disrepair. Whether in celebration or mourning, these events echo a truth no architect can outrun: joy collapses fastest when the warnings are built into the walls.

Sources (*and related articles*):

- *Wikipedia.* "Estádio Fonte Nova." https://en.wikipedia.org/wiki/Est%C3%A1dio_Fonte_Nova
- *Wikipedia* (Português). "Estádio Octávio Mangabeira" https://pt.wikipedia.org/wiki/Est%C3%A1dio_Oct%C3%A1vio_Mangabeira
- *Reuters.* "Seven dead in Brazilian stadium disaster." https://www.reuters.com/article/sports/seven-dead-in-brazilian-stadium-disaster-idUSN25287040/
- *UOL Esporte.* "Arquibancada da Fonte Nova cede e deixa vítimas – 25/11/2007." (Português).

https://www.uol.com.br/esporte/futebol/ultimas/2007/11/25/ult59u138101.jhtm

- *GloboEsporte.com*. "Tragédia da Fonte Nova completa dez anos" (revisita o incidente com informações e testemunhos). (Português). https://ge.globo.com/ba/noticia/abriu-um-clarao-10-anos-depois-tragedia-da-fonte-nova-contada-por-quem-estava-la.ghtml

27.
Sayano-Shushenskaya Dam Disaster (Russia, 2009):
When Turbines Betray the Power They Produce

On the morning of 17 August 2009, the world's sixth-largest hydroelectric power station—Sayano–Shushenskaya, perched on the Yenisei River in Siberia—roared into motion as usual. Designed to generate a staggering 6,400 MW, it powered millions across Russia. But beneath the turbines lurked a ticking overload. Over decades, vibrations grew unchecked, metal fatigue spread, and crucial sensors were ignored or deactivated to avoid shutdowns. In one fatal decision, operators bypassed alarms designed to detect turbine imbalance.

At 08:13, turbine number 2 catastrophically disintegrated—its massive rotor broke apart, ripping through the cavern with a force comparable to hundreds of explosive charges. Water surged back into the station, generating an uncontrollable flood that tore through the turbine hall, destroyed the dam's powerhouse, and sent murky torrents of oil and debris downstream.

The human toll was devastating: 75 workers lost their lives, and over a hundred were injured. The station was left gutted, its electrical network crippled, and thousands were plunged into blackout. The disaster was attributed not to natural calamity, but to a cascade of engineering failures: turbine subunits that exceeded safe operating time, inadequate vibration dampening, and structural fatigue that went unresolved. Russia's supervisory agencies later admitted they had overlooked sensor logs and inspection protocols in the name of maximizing power output.

In the shadow of the towering dam, a chilling lesson was learned: power systems are only as strong as the diligence behind their maintenance. Similar mechanical betrayals have claimed lives elsewhere—whether in derailed trains, collapsed industrial floors, or blinded safety warnings. At Sayano, the turbines didn't just fail—they took 75 souls with them in an echoing void.

Sources (*and related links*):

- *Wikipedia.* "Sayano–Shushenskaya hydroelectric power station accident." https://en.wikipedia.org/wiki/Sayano–Shushenskaya_hydroelectric_power_station_accident

- *PBS NOVA.* "Turbine Failure at Sayano–Shushenskaya Dam." https://www.pbs.org/wgbh/nova/article/turbine-failure-sayano-shushenskaya-dam/

- *World Nuclear News.* "Investigation: Sayano–Shushenskaya accident." https://www.world-nuclear-news.org/Articles/Investigation-Sayano-Shushenskaya-accident

- *The Guardian.* "Russia's hydroelectric power station disaster kills dozens." https://www.theguardian.com/world/2009/aug/17/russia-hydroelectric-dam-accident

28.
Korba Chimney Collapse (India, 2009):
A Tower of Neglect

Amid torrential rain and crackling thunder on September 23, 2009, a towering chimney under construction at the Bharat Aluminium Company (BALCO) power plant in Korba, Chhattisgarh, came crashing down. The 275-meter concrete stack—meant to be one of the tallest in India—collapsed in seconds, burying dozens of workers beneath thousands of tons of debris. At least 45 lives were lost in the disaster, many of them daily wage laborers who had little protection and nowhere to escape.

Though the collapse occurred during a thunderstorm, the failure was far from an act of nature alone. Investigations revealed a series of deeply troubling lapses: substandard construction practices, poor-quality materials, and a complete lack of engineering supervision on-site. The contractor was found to have deviated from approved design plans and ignored safety protocols. Shockingly, work on the structure had continued despite official weather warnings—putting profits ahead of people. Eyewitnesses described rushed timelines, inadequate scaffolding, and the absence of routine inspections as part of a broader culture of negligence.

The Korba tragedy sent shockwaves through India's infrastructure sector, which was then expanding rapidly to meet growing energy demands. It exposed the vulnerability of laborers at high-risk construction sites and highlighted a dangerous tendency among contractors and clients to prioritize deadlines over safety. In the wake of the collapse, there were calls for stricter enforcement of building codes, improved site inspections, and legal accountability for developers. But for the families of the workers who died, these reforms came too late. The fallen chimney stood not as a monument to progress, but as a grim reminder of what happens when ambition rises faster than responsibility.

The Korba chimney did not fall in silence—it echoed across time zones and continents where safety was equally sidelined. In China's Fengcheng power plant collapse (2016), 74 workers died when a cooling tower under construction gave way due to rushed timelines and ignored warnings. In Bangladesh's Rana Plaza disaster (2013), structural failure claimed over 1,100 lives in a building never meant to carry the weight it bore. And in Cambodia's Sihanoukville building collapse (2019), 28 construction workers perished under layers of corruption, concrete, and complacency. These incidents serve as harsh footnotes to Korba's fall—each a reminder that height means nothing if the foundations of ethics, oversight, and accountability are hollow.

Sources (*and related articles*):

- *Wikipedia.* "2009 Korba chimney collapse." https://en.wikipedia.org/wiki/2009_Korba_chimney_collapse
- *Global Energy Monitor.* "2009 Korba chimney collapse." https://www.gem.wiki/2009_Korba_chimney_collapse

- *POWER Magazine.* "Death Toll at Indian Power Plant Chimney Collapse Rises to 46." https://www.powermag.com/death-toll-at-indian-power-plant-chimney-collapse-rises-to-46/
- *Times of India.* "2009 chimney collapse case: Probe indicts Balco, Chinese firm & others for negligence." https://timesofindia.indiatimes.com/business/india-business/2009-chimney-collapse-case-probe-indicts-balco-chinese-firm-others-for-negligence/articleshow/17705992.cms
- *France24.* "At least 20 dead in power plant accident." https://www.france24.com/en/20090923-least-20-dead-power-plant-accident-

29. Fukushima Daiichi Nuclear Disaster (Japan, 2011):

A Triple Disaster Unleashes a Nuclear Crisis

On March 11, 2011, a massive 9.0-magnitude undersea earthquake off Japan's northeastern coast triggered a devastating tsunami that swept over coastal towns and surged into the Fukushima Daiichi Nuclear Power Plant. Operated by Tokyo Electric Power Company (TEPCO), the facility lost all external power, and the tsunami disabled emergency diesel generators. With cooling systems offline, three reactor cores overheated, resulting in partial meltdowns, hydrogen explosions, and the release of radioactive material into the atmosphere and the Pacific Ocean.

The disaster prompted the evacuation of over 150,000 residents and raised global alarm over nuclear safety. Although no immediate deaths were attributed directly to radiation exposure, long-term health effects, mental health impacts, and disruptions to livelihoods and ecosystems were severe. Investigations pointed to multiple failures: insufficient tsunami defenses, poor crisis preparedness, and regulatory complacency. The crisis also exposed communication breakdowns between TEPCO, the Japanese government, and the public, contributing to panic and eroding trust.

In the aftermath, Japan temporarily shut down all its nuclear reactors and overhauled its nuclear safety framework.

Internationally, the event reignited debates over nuclear energy's role in a disaster-prone world. More than a decade later, Fukushima remains a symbol of what happens when natural disasters collide with technological overconfidence.

From Chernobyl to Fukushima, the cost of nuclear miscalculation is measured not just in radioactive fallout but in the collapse of public trust. When emergency systems fail and contingency plans prove hollow, even a facility engineered for resilience can become ground zero for a cascading catastrophe.

Sources (*and related articles*):

- *Wikipedia*. "Fukushima Daiichi nuclear disaster" https://en.wikipedia.org/wiki/Fukushima_Daiichi_nuclear_disaster
- *World Nuclear Association*. "Fukushima Daiichi Accident" https://world-nuclear.org/information-library/safety-and-security/safety-of-plants/fukushima-daiichi-accident.aspx
- *IAEA (International Atomic Energy Agency)*. "Fukushima Daiichi Nuclear Accident." https://www.iaea.org/topics/response/fukushima-daiichi-nuclear-accident
- *UNSCEAR (United Nations Scientific Committee on the Effects of Atomic Radiation)*. "The Fukushima-Daiichi Nuclear Power Station Accident: An overview." https://www.unscear.org/unscear/en/fukushima.html
- *ScienceDirect* (abstract). "Endogenous causes, risk perception and policy responses to the Fukushima crisis." https://www.sciencedirect.com/science/article/abs/pii/S2212420923001309

30.
Once Station Train Disaster (Argentina, 2012):
When Momentum Turns Fatal

In the early hours of a typical Wednesday morning, on Feb 22, 2012, commuters poured into the bustling Once Station in Buenos Aires, unaware they were about to witness one of Argentina's worst rail disasters in decades. The eight-carriage train—a Sarmiento Line commuter service—pulled into the terminal at nearly 50 km/h, nearly double the safe approach speed. The motorman reported brake issues, but the train pressed on. At 08:33 a.m., it slammed into the buffer stops at a speed of 26 km/h. The impact telescoped the first two carriages, crumpling the lead coach into the platform with devastating effect.

The casualty numbers are heartbreaking: 51 people were killed and over 700 injured, many suffering life-altering trauma. Rescue teams had to cut open roofs to reach trapped passengers. Photos captured commuters strewn across the platform, immobilized beneath twisted metal, as ambulances and helicopters raced the injured to hospitals.

Engineer reports later found a lethal mix of brake failure, excessive speed, and systemic neglect. The train's operator, Trenes de Buenos Aires (TBA)—run by the Cirigliano brothers—had skimped on maintenance amid allegations of corruption and misuse of government subsidies. The national government intervened, revoked TBA's contract, and launched historic prosecutions for corporate and political figures

Though the station has since reopened and the scars have begun to heal, the Once disaster remains a chilling testament to the consequences of compromised infrastructure, cost-cutting, and regulatory failure. It forced Argentina—and countless other nations—to face hard questions: how safe is "routine," and what price do we pay when oversight fails at critical junctions? Similar tragedies echo this theme across continents: the Pukhrayan derailment in India (2016), where track fatigue and maintenance delays killed over 150; the Washington Metro crash (2009), linked to aging control systems; and the Santiago derailment in Spain (2013), where speed, oversight gaps, and inadequate signage proved fatal. Each case offers a sobering reminder: when safety takes a backseat, the cost is counted in lives.

Sources (*and related articles*):

- *Wikipedia.* "2012 Buenos Aires rail disaster." https://en.wikipedia.org/wiki/2012_Buenos_Aires_rail_disaster
- *Reuters.* "Argentine commuter train crashes, killing 49 people." https://www.reuters.com/article/markets/argentine-commuter-train-crashes-killing-49-people-idUSL2E8DM5F2/
- *The Guardian.* "At least 49 dead and hundreds injured as train crashes in Argentina." https://www.theguardian.com/world/2012/feb/22/argentinian-commuters-injured-train-crash
- *El País.* "Medio centenar de muertos en un accidente ferroviario en Argentina."

https://elpais.com/internacional/2012/02/22/actualidad/1329921646_764629.html

31.
Thane Building Collapse (India, 2013):
The Weight of Unauthorized Floors

In the spring of 2013, a towering structure in *Shil Phata*, a developing suburb of Thane near Mumbai, buckled under its own illegitimacy. What stood as a seven-story residential building collapsed without warning on April 4, reducing entire families to rubble. In the end, 74 people lost their lives, including children, women, and laborers—many of whom had just moved in or were still helping with construction on the upper floors.

The tragedy wasn't rooted in engineering miscalculation so much as deliberate evasion of law and oversight. The building, officially unauthorized, had been constructed without approval, on land not zoned for residential use, and with materials and workmanship that fell far short of any reasonable safety standard. It had risen rapidly, floor after floor, often overnight—a common symptom of the region's unchecked urban sprawl. Alarmingly, while only four floors were originally planned, three additional floors had been added illegally, even as tenants were being allowed to move in.

Investigations revealed complicity across the board—from corrupt municipal officials turning a blind eye, to local developers exploiting housing demand. The floors lacked proper columns and reinforcements, and basic seismic and structural considerations were ignored. As one resident later said, "We moved in with dreams. We had no idea we were moving into a tomb."

The Thane collapse triggered public outrage and a brief crackdown on unauthorized construction across Maharashtra. A few arrests followed, including that of the builder and some civic staff, but systemic reform remained elusive. Even today, in India's densely populated metros, illegal high-rises continue to dot the skyline—monuments to ambition built on fragile foundations.

Similar tragedies have echoed in cities grappling with lax enforcement and high housing pressure. In Nairobi's Huruma district (Kenya, 2016), a six-story building, expanded illegally, crumbled under floodwaters. In Phnom Penh (Cambodia, 2019), a seven-story construction site collapsed during unauthorized nighttime work, killing 28 workers. And in Alexandria (Egypt, 2008), a 12-story residential building fell like a stack of cards after being illegally extended—an enduring symbol of ignored warnings and fragile oversight.

Sources (and related articles):

- *Wikipedia.* "2013 Thane building collapse." https://en.wikipedia.org/wiki/2013_Thane_building_collapse
- *The Guardian.* "Mumbai building collapse kills dozens." https://www.theguardian.com/world/2013/apr/05/mumbai-building-collapse-kills-dozens
- *BBC News.* "India building collapse: Dozens dead in Mumbai suburb." https://www.bbc.com/news/world-asia-india-22038224

- *NDTV*. "Illegal Building Collapses in Thane, 74 Dead." https://www.ndtv.com/india-news/illegal-building-collapses-in-thane-74-dead-517409

32.
Rana Plaza Collapse (Bangladesh, 2013):
Buried Beneath the Seams

April 24, 2013, began like any other workday in Savar, a suburb of Dhaka, Bangladesh. But by mid-morning, the eight-story Rana Plaza building crumbled into a mountain of dust, concrete, and tangled steel. In less than 90 seconds, over 1,130 garment workers were dead, and more than 2,500 others were injured—many crushed, many trapped for days in the rubble. It remains one of the deadliest structural failures in modern history, and a defining tragedy in the global garment industry.

What made the collapse, so horrifying was not just the scale, but the preventability. Rana Plaza was never meant to house factories. Originally constructed as a modest commercial complex, it had been illegally expanded by additional floors to accommodate heavy textile machinery the structure was never designed to bear. The concrete was weak, the columns thin, and the foundation wholly inadequate. Cracks had appeared throughout the building a day before the disaster—large enough to alarm television crews and structural engineers—but the building's owners insisted it was safe. Workers were ordered back inside under threat of withheld wages.

The tragedy didn't just expose fatal flaws in a single building—it ripped open the global conscience. It laid bare a system in which lax enforcement, corruption, and relentless pressure for cheap labor collided with devastating results. The collapse forced international retailers to reckon with the hidden cost of fast fashion, triggering widespread protests, boycotts, and eventually, the creation of legally binding safety accords for Bangladeshi factories. But for thousands of families, no global reform could undo the personal loss. Rana Plaza became a symbol not only of engineering failure, but of a global supply chain built on fragile foundations.

Rana Plaza did not fall because of one miscalculation. It collapsed under the weight of greed, coercion, and the global silence that had long accompanied unsafe labor practices. Its echoes have been heard elsewhere: in Karachi (Pakistan, 2012), where a garment factory fire killed 260 workers locked inside; in Delhi (India, 1997), where an illegally expanded cinema collapsed, killing over 50; and in Tangshan (China, 2020), where a hotel used for COVID-19 quarantine fell due to faulty modifications, claiming dozens of lives. Each disaster is a thread in the same fabric, revealing a global pattern where the price of speed and profit is measured not in dollars, takas, rupees, or yuans, but in human lives buried beneath concrete and compromise.

Sources (*and related articles*):

- *Wikipedia.* "Rana Plaza collapse." https://en.wikipedia.org/wiki/Rana_Plaza_collapse
- *BBC News.* "Bangladesh Building Collapse: Death Toll Passes 1,000." *BBC*, 13 May 2013. https://www.bbc.com/news/world-asia-22476774

- *Reuters.* "Six years after Bangladesh's Rana Plaza disaster, fashion brands urged to pay more." https://www.reuters.com/article/world/six-years-after-bangladeshs-rana-plaza-disaster-fashion-brands-urged-to-pay-mo-idUSKBN1HI0XS/
- *Time.* "Bangladesh Factory Owner Charged With Murder Over Rana Plaza Collapse." https://time.com/3902945/rana-plaza-murder-charges-owner-sohel-rana/
- *Congressional Research Service.* "Bangladesh Apparel Factory Collapse: Background in Brief." https://sgp.fas.org/crs/row/R43085.pdf

33.
Tongaat Mall Collapse (South Africa, 2013):
Warnings Buried in Concrete

Construction workers were still laying brick and pouring concrete at the Tongaat Mall site near Durban when, without warning, a section of the partially built structure gave way on November 19, 2013. Two workers were killed instantly, and 29 others were seriously injured. The collapse shocked the nation—not because it happened suddenly, but because it happened after repeated warnings had already been issued.

Authorities had served multiple stop-work notices on the developer prior to the incident, citing visible structural irregularities and unsafe building practices. These notices were ignored. Investigators later found that construction had deviated from the approved engineering plans, with unauthorized modifications and substandard concrete pouring compromising the building's integrity. Foundations were inadequately supported, and worst of all, no certified structural engineer had signed off on the reinforcements at the time of the collapse.

The tragedy not only exposed technical failure—it highlighted a dangerous culture of disregard for regulatory oversight in South Africa's commercial construction sector. While legal proceedings were launched against both the developer and the contractor, the case dragged on for years, further eroding public confidence in accountability. What made this collapse especially haunting was not what engineers didn't know—but what they knew and were ignored for. When red flags are buried beneath fresh concrete, disaster isn't a question of if—it's a matter of when.

The fall of the Tongaat Mall was less an engineering mystery and more a failure of enforcement—proof that when warnings are silenced and oversight sidelined, concrete becomes a coffin. Similar tragedies have emerged elsewhere: in Mexico City (1985), where school buildings collapsed after ignored retrofitting orders; in Gwangju (South Korea, 2021), where an apartment demolition gone wrong crushed a bus after safety protocols were skipped; and in Nairobi (Kenya, 2015), where stop-work notices were routinely flouted before a six-story building gave way. These failures share a common blueprint: one where the loudest voice isn't that of the engineer—but of the one rushing to finish, no matter the cost.

Sources (*and related articles*):

- *Al Jazeera.* "Many trapped in S Africa mall collapse." https://www.aljazeera.com/news/2013/11/20/many-trapped-in-s-africa-mall-collapse
- *France 24.* "Deadly South African mall collapse traps dozens." https://www.france24.com/en/20131119-south-african-mall-collapse-kills-one-traps-50-say-officials
- *IOL* (Independent Online). "PICS: Deadly Tongaat mall collapse." https://iol.co.za/news/south-africa/kwazulu-natal/2013-11-19-pics-deadly-tongaat-mall-collapse/

34.
Dharahara Tower Collapse (Nepal, 2015):
Heritage Humbled by Neglect

In the heart of Kathmandu, Nepal's capital, the iconic Dharahara Tower stood tall for nearly two centuries—a symbol of national pride, resilience, and architectural legacy. But on April 25, 2015, when a powerful 7.8-magnitude earthquake rocked the region, that symbol came crashing down. The historic tower crumbled into dust and rubble, burying dozens of visitors beneath it. Over 40 people lost their lives in the collapse, many of them local residents and tourists who had climbed its spiral staircase for a view of the city skyline just moments before.

Originally constructed in 1832 during the reign of Prime Minister Bhimsen Thapa, the Dharahara had weathered time, conflict, and even previous earthquakes. Though damaged in the 1934 quake and rebuilt afterward, the tower remained largely untouched by modern engineering standards. Its load-bearing brick masonry and lack of internal reinforcement made it inherently vulnerable to seismic shocks. Engineers and heritage conservationists had raised concerns about its structural fragility and the urgent need for retrofitting, but preservation plans remained bogged down in bureaucracy and underfunding.

The collapse of the Dharahara became more than a headline—it became a sobering metaphor for Nepal's broader infrastructural fragility. As the nation grappled with thousands of casualties, destroyed homes, and collapsed temples, the fall of this national monument reminded the world that heritage must be preserved not just in spirit, but in structure. Cultural icons, no matter how beloved, cannot be immune to time and tectonics. When symbolic strength isn't matched by physical resilience, the past—and the people visiting it—can be lost in an instant.

The fall of the Dharahara was not just the loss of a tower—it was the fall of memory, stone by stone. Similar tragedies have shaken history elsewhere: the Qutb Minar complex in Delhi has faced tremors and weathering with only partial reinforcement; Bam's 2,000-year-old citadel in Iran crumbled almost entirely in the 2003 earthquake; and the Basilica of St. Benedict in Norcia, Italy, collapsed during the 2016 quake, despite its sacred status. These structures stood as heritage—but they fell as reminders: preserving the past demands more than admiration; it requires engineering action, before reverence turns into ruin.

Sources (*and related articles*):

- *BBC News.* "Nepal Earthquake: Iconic Dharahara Tower in Kathmandu Collapses." *BBC*, 25 Apr 2015. https://www.bbc.com/news/world-asia-32461019

- *Wikipedia.* "Dharahara." https://en.wikipedia.org/wiki/Dharahara

- *The Guardian.* "Nepal quake kills more than 1800 and spreads terror on Everest."

https://www.theguardian.com/world/2015/apr/25/nepal-quake-kills-more-than-1000

- *NDTV*. "Historic Dharahara Tower Collapses in Kathmandu After Earthquake, 180 Killed." https://www.ndtv.com/india-news/historic-dharahara-tower-collapses-in-kathmandu-after-earthquake-180-killed-758133

- *Britannica*. "Nepal earthquake of 2015." https://www.britannica.com/topic/Nepal-earthquake-of-2015

35.
Tainan Earthquake Building Collapse (Taiwan, 2016):
Cooking Oil Cans Where Steel Should've Been

In the early hours of February 6, 2016, a 6.4 magnitude earthquake struck southern Taiwan, shaking the city of Tainan and jolting thousands from their sleep. Among the worst-hit structures was the Weiguan Jinlong residential complex, a 17-story high-rise that pancaked into itself with terrifying speed. Within seconds, entire families were buried beneath tons of concrete and twisted steel. Rescue workers labored day and night for nearly a week, pulling survivors from the rubble and recovering the bodies of those who didn't make it. In total, 115 people lost their lives in the collapse, making it the deadliest outcome of the quake.

What stood out most about the Weiguan disaster wasn't just the scale of the destruction—it was the contrast. Other buildings in the immediate vicinity, exposed to the same seismic forces, remained upright or suffered only minor damage. Questions mounted quickly, and so did suspicions. As engineers and investigators sifted through the debris, they uncovered a disturbing truth: large, empty cooking oil cans had been embedded within the walls of the building during its construction. These were used in place of proper concrete fill and steel reinforcement—likely as a cost-cutting measure. The result was a catastrophic compromise in structural integrity, hidden beneath layers of paint and plaster.

The building, constructed in the early 1990s during Taiwan's real estate boom, had passed years of inspections without raising alarms. But the quake revealed the consequences of short-term savings and lax oversight. Public outrage followed swiftly, and the contractor responsible was arrested for gross negligence and violation of building codes. The tragedy became a national wake-up call, prompting widespread inspections of older high-rise buildings across the country. New policies were introduced to tighten seismic safety standards and to hold developers accountable—not just for new builds, but for the legacy of what had already been constructed.

The fall of the Weiguan Jinlong tower wasn't merely the result of an earthquake—it was the collapse of trust in a system meant to safeguard lives. It showed that in seismic zones, corners cut decades ago can come back with lethal force, and that behind every structural decision lies an unspoken contract with the future.

The Weiguan Jinlong collapse revealed more than faulty design—it exposed how corruption and complacency can embed disaster into a city's very foundations. Taiwan's tragedy echoed past catastrophes: in Turkey's 1999 İzmit earthquake, entire blocks of "pancake buildings" fell due to substandard concrete and weak columns; during Italy's 2009 L'Aquila quake, retrofitted structures still crumbled under falsified safety reports; and in Mexico City's 2017 quake, schools and homes collapsed where bribes had replaced blueprints. In each case, it wasn't the earth alone that killed—it was what had been buried inside the walls.

Sources (*and related articles*):

- *BBC News*. "Deadly earthquake topples buildings in Taiwan city of Tainan." *BBC*, 6 Feb 2016. https://www.bbc.com/news/world-asia-35508475
- *CNN*. "Taiwan Earthquake: Rescuers Dig for Survivors in Collapsed Building." *CNN*, 7 Feb 2016. https://www.cnn.com/2016/02/06/asia/taiwan-earthquake/index.html
- *Wikipedia*. "2016 southern Taiwan earthquake." https://en.wikipedia.org/wiki/2016_southern_Taiwan_earthquake
- *The Guardian*. "Taiwan earthquake: developer of collapsed tower block arrested." https://www.theguardian.com/world/2016/feb/09/taiwan-earthquake-developer-of-collapsed-tower-block-arrested
- *Time*. "34 Dead, More Than 100 Still Trapped After Earthquake Hits Taiwan." https://time.com/4211109/taiwan-earthquake/
- *NOAA NCEI Hazard Database* – USGS record of the quake https://www.ngdc.noaa.gov/hazel/view/hazards/earthquake/event-more-info/10174

36.
Nairobi Building Collapse (Kenya, 2016):
Rainfall and Rotten Concrete

Heavy rains had been lashing Nairobi in the final days of April 2016 when tragedy struck in the densely populated Huruma neighborhood. A six-story residential building, already teetering on structural failure, gave way and crumbled into a heap of concrete and dust. At least 51 people lost their lives in the collapse, with many more injured or buried beneath the debris. Families were torn apart in an instant, and emergency crews scrambled for days to recover survivors.

What made the incident even more harrowing was the fact that this was no unforeseen disaster. The building had already been condemned and tagged for demolition by local authorities—but it remained fully occupied. Residents, many of them low-income earners, had little choice but to stay. The structure itself stood on unstable, marshy ground and lacked a proper foundation. Investigations later revealed the use of substandard materials—weak concrete, poorly tied rebar—and, perhaps most damning, the illegal addition of extra floors beyond the approved plan.

The Huruma collapse served as a grim symbol of a broader crisis—one where the pace of urban growth outstrips the enforcement of safety regulations. In many rapidly expanding African cities, unchecked construction, weak governance, and housing demand collide in dangerous ways. This disaster was not just about one faulty building; it was about a system stretched too thin, where oversight failed and lives were the cost.

The Huruma tragedy echoed a chilling pattern seen across the Global South—where housing booms are often built on brittle promises. From the Maputo courthouse collapse in Mozambique to Lagos's frequent construction failures in Nigeria, and Cairo's illegally extended flats, the story repeats: concrete poured faster than caution, foundations laid thinner than the law. In Nairobi, it wasn't just a building that collapsed—it was trust in a system that allowed condemned structures to remain occupied. And until construction is rooted in accountability, every rainfall may test not just roofs, but resolve.

Sources (*and related links*):

- *Wikipedia.* "Huruma" (Nairobi estate including collapse details)
  https://en.wikipedia.org/wiki/Huruma
- *Deutsche Welle.* "Nairobi six-storey building collapses"
  https://p.dw.com/p/1IfZY
- *Reuters.* "Kenya charges six men over building collapse that killed 51"
  https://www.reuters.com/article/world/kenya-charges-six-men-over-building-collapse-that-killed-51-idUSKCN0YU0JR/

37.
Kintampo Waterfall Tragedy (Ghana, 2017):
Nature Struck, but Safety Failed

A weekend visit to one of Ghana's most scenic natural attractions turned deadly on March 19, 2017, when a sudden storm unleashed disaster at the Kintampo Waterfalls in the Brong-Ahafo Region. As dozens of students from a local school enjoyed the cascading waters and took photos on a viewing platform, powerful winds toppled several massive trees perched above the site. In a matter of moments, the trees crashed down onto the crowd, killing 20 people—most of them teenagers—and injuring many others.

While the tragedy was largely driven by natural forces, subsequent investigations exposed critical man-made failures that amplified the loss of life. The viewing platform, constructed to enhance the visitor experience, had been built without proper safety features. There were no reinforced railings, no designated evacuation paths, and no integration of basic weather monitoring systems—despite the site's growing popularity as a tourist destination. Local authorities had failed to implement even minimal environmental risk assessments, and emergency preparedness was virtually nonexistent.

The Kintampo tragedy underscored a hard truth: even in natural settings, the presence of human infrastructure brings with it a duty of care. Engineering is not only about bridges, dams, and towers—it's also about how we interface with nature, especially in spaces where visitors gather in large numbers. This disaster served as a wake-up call that safety planning, environmental awareness, and structural responsibility must extend beyond the built environment—because when nature lashes out, poorly designed spaces offer little protection.

The Kintampo tragedy joined a growing list of disasters where human oversight faltered at nature's edge. Similar events—from the Vajont Dam overtopping in Italy, where a mountainside's wrath was underestimated, to Taiwan's collapsing riverwalk canopies—reveal how even scenic environments demand engineering foresight. In Ghana, the absence of risk planning turned a place of beauty into one of mourning. It wasn't the trees alone that fell—it was the illusion that nature, left unmanaged, won't test the limits of human design.

Sources (*and other related links*):

- *Wikipedia.* "Kintampo waterfalls." https://en.wikipedia.org/wiki/Kintampo_waterfalls
- *The Guardian.* "Twenty teenagers die in Ghana waterfall accident." https://www.theguardian.com/world/2017/mar/20/twenty-teenagers-die-ghana-waterfall-accident-kintampo
- *Deutsche Welle.* "Freak accident at Ghana waterfall kills several – DW – 03/20/2017." https://www.dw.com/en/freak-accident-at-ghana-waterfall-kills-several-students/a-38026881

- *UPI.* "20 dead, mostly students, after trees fall in Ghana's Kintampo waterfalls." https://www.upi.com/Top_News/World-News/2017/03/20/20-dead-mostly-students-after-trees-fall-in-Ghanas-Kintampo-waterfalls/9741490008430/

38. Dakar Stadium Stampede Structure Failure (Senegal, 2017): *Cracks Beneath the Roar*

On the night of July 15, 2017, what began as a high-energy football match at Dakar's Demba Diop Stadium descended into tragedy. During the final whistle of a tightly contested game, tensions flared between rival fans. As chaos broke out in the stands, a stampede erupted near one of the stadium's exits. Amid the panic and crowd surge, a section of the outer wall gave way, crumbling under the sudden weight of fleeing bodies. Eight people lost their lives, and more than sixty were injured.

In the aftermath, investigators found that the wall had long been compromised. Years of exposure to Senegal's coastal climate had eroded the concrete, while poor maintenance and lack of structural reinforcement had left it brittle and vulnerable. Worse still, the stadium had been filled beyond its safe capacity, and the exits were poorly designed for emergency egress. As hundreds pressed toward a narrow escape route, the accumulated pressure proved too much for a wall never meant to absorb such dynamic load. There were no effective crowd control measures in place, no visible emergency response, and no preventive inspections that could have averted the collapse.

Public outcry followed swiftly, not just over the loss of life, but over the conditions that had made it possible. The tragedy sparked a nationwide review of stadium safety standards and long-overdue audits of aging sports infrastructure throughout Senegal. Yet, for the families of those lost, the damage was irreversible. The collapse at Demba Diop served as a grim reminder that structural decay is not always visible—and that the passion of a crowd, when met with poor planning, can turn deadly in seconds. Beneath the roar of celebration, cracks had been forming for years. No one looked until it was too late.

The Demba Diop tragedy echoed similar calamities worldwide—from the 2001 Ellis Park stampede in South Africa to the 2022 Kanjuruhan Stadium disaster in Indonesia—where structural vulnerabilities met unrestrained crowds. These events underscore a sobering truth: stadiums are not just arenas of emotion; they are pressure vessels of human movement. Without structural integrity, controlled egress, and crowd management, even moments of celebration can collapse into chaos. In Dakar, it wasn't just concrete that crumbled—it was the trust that infrastructure would hold when it mattered most.

Sources (*and related articles*):

- *Wikipedia.* "Demba Diop stadium crush." https://en.wikipedia.org/wiki/Demba_Diop_stadium_crush
- *Al Jazeera.* "Eight dead in Senegal football stadium stampede." https://www.aljazeera.com/sports/2017/7/16/eight-dead-in-senegal-football-stadium-stampede

- *Euronews.* "Eight dead in Senegal after stampede at football league final." https://www.euronews.com/2017/07/16/eight-dead-in-senegal-after-stampede-at-football-league-final

39.
Genoa Morandi Bridge Collapse (Italy, 2018):
The Weight of Unspoken Warnings

On August 14, 2018, during a heavy summer downpour, a massive section of the Morandi Bridge in Genoa, Italy, gave way without mercy. In mere seconds, 200 meters of roadway collapsed, sending dozens of vehicles plunging more than 45 meters into the void. Forty-three lives were lost.

The bridge—an ambitious cable-stayed structure opened in 1967—had long been lauded as a marvel of mid-century engineering. Yet behind its sweeping silhouette lay a tragic vulnerability: corrosion silently gnawed at the steel tendons encased within its concrete stays, weakening its very soul. Years of deferred maintenance, missed warnings, and bureaucratic lethargy compounded the danger. Engineers had raised red flags; academics had urged action. But crucial retrofitting was either delayed or dismissed.

The collapse was more than a structural failure—it was an indictment. An indictment of systemic complacency in Italy's infrastructure governance, especially where private operators like *Autostrade per l'Italia* were responsible for toll roads but opaque in their maintenance regimes. The disaster triggered public outrage, legal investigations, and a national reckoning over infrastructure transparency and accountability.

The Morandi Bridge's fall became not only a symbol of crumbling concrete but of crumbling trust.

The Genoa tragedy joined a global chorus of catastrophic bridge failures—Minneapolis in 2007, Kolkata's Majerhat Bridge in 2018, and Taiwan's Nanfang'ao in 2019—where structural aging, ignored warnings, or rushed repairs played fatal roles. The Morandi Bridge collapse wasn't merely about physics or design—it was about the danger of silence in the face of known risk. When routine inspections become rituals, and warnings gather dust, even the mightiest spans can break under the weight of inaction. Genoa reminded the world that infrastructure doesn't just carry traffic—it carries trust.

Sources (*and related articles*):

- *Wikipedia.* "Ponte Morandi Collapse." *Wikipedia*, last modified 2024. https://en.wikipedia.org/wiki/Ponte_Morandi_collapse
- *VOA News.* "59 on Trial in 2018 Italy Bridge Collapse That Killed 43." *VOA*, 7 July 2022. https://www.voanews.com/a/on-trial-in-2018-italy-bridge-collapse-that-killed-43-/6649699.html
- *ProMarket.* "Two Years Ago, the Morandi Bridge Collapse Claimed 43 Lives." *Stigler Center*, 30 June 2020. https://www.promarket.org/2020/06/30/two-years-ago-the-morandi-bridge-collapse-claimed-43-lives-a-new-stigler-center-case-study-explores-the-link-between-the-disaster-and-crony-capitalism/

- *Designing Buildings Wiki*. "Genoa Bridge Collapse." *Designing Buildings*, 2018. https://www.designingbuildings.co.uk/wiki/Genoa_bridge_collapse
- *AP News*. "Italy Marks 5th Anniversary of Morandi Bridge Collapse with Demands for Justice for the 43 Killed." *AP News*, 14 Aug 2023. https://apnews.com/article/italy-morandi-bridge-collapse-c1c2864ac02135429534a7ebdcbe1beb
- *Reuters*. "Italy's Anti-Establishment Government Angers Creditors over Bridge Collapse Costs." *Reuters*, 15 Aug 2018. https://www.reuters.com/article/us-italy-bridge-collapse-idUSKBN1L0273

40.
Brumadinho Dam Disaster (Brazil, 2019):
Where Iron Flowed and Truth Sank

On January 25, 2019, a routine afternoon at the Córrego do Feijão iron ore mine gave way to horror. Without warning, Dam I—an upstream tailings structure operated by mining giant Vale S.A.—ruptured, unleashing a violent surge of over 11 million cubic meters of mining sludge. The deluge tore through administrative offices, a bustling lunchroom, worker dorms, and nearby settlements. At least 270 people were killed—many entombed in a wave of iron-rich mud.

The rupture was not merely mechanical. It was institutional.

Only months before, the dam had received safety certification. Yet post-disaster investigations unearthed a deeply unsettling truth: internal documents, falsified stability declarations, and suppressed data painted a picture of willful negligence. Sensor anomalies, known drainage issues, and warnings from geotechnical experts had been buried—just like the victims—in favor of economic expedience. The upstream construction method used, though cheaper, was inherently unstable, especially in Brazil's wet season.

Brumadinho wasn't an anomaly; it was an echo. Just three years earlier, in 2015, the Mariana dam collapse had signaled a similar failure. But little had changed. This second tragedy ignited a national reckoning—sparking mass protests, criminal prosecutions, multi-billion-dollar lawsuits, and an overhaul of Brazil's mining oversight protocols.

The scar on the land remains. But deeper still is the erosion of trust—between people, regulators, and an industry too long allowed to police itself.

It was not Brazil's first tailings tragedy—just the most unforgivable. Less than four years earlier, Mariana had warned what could come when containment fails. Brumadinho was not an accident—it was an aftershock.

Sources (*and related articles*):

- *Wikipedia.* "Brumadinho Dam Disaster." *Wikipedia*, last modified 2024. https://en.wikipedia.org/wiki/Brumadinho_dam_disaster
- *Reuters.* "Brazil's Vale fined $17 mln for Brumadinho tailings dam disaster." *Reuters*, 15 Aug 2022. https://www.reuters.com/business/brazils-vale-fined-17-mln-brumadinho-tailings-dam-disaster-2022-08-15/
- *ETH Zürich.* "Brumadinho dam collapse: The danger emerged after the decommissioning." *ETH Zürich*, Jan 2022. https://www.ethz.ch/en/news-and-events/eth-news/news/2024/01/brumadinho-dam-collapse-the-danger-emerged-after-the-decommissioning.html
- *Public Eye.* "Vale and the Brumadinho dam collapse." *Public Eye*, 2019. https://www.publiceye.ch/en/topics/commodities-trading/rohma/when-commodity-traders-get-caught/vale-and-the-brumadinho-dam-collapse

- *Mongabay*. "Scientists now know how the Brumadinho dam disaster happened …" *Mongabay*, Feb 2024. https://news.mongabay.com/2024/02/scientists-now-know-how-the-brumadinho-dam-disaster-happened-and-the-lessons-to-learn/

41.
Phnom Penh Building Collapse (Cambodia, 2019):
The Cost of Rising Too Fast

In the pre-dawn stillness of June 22, 2019, a seven-story building under construction in Sihanoukville crumbled without warning, burying dozens beneath the weight of greed and negligence. Twenty-eight workers were killed, many in their sleep. They had been living inside the unfinished structure—a grim necessity born of poor wages and nonexistent worker accommodations.

The building was part of a frenzied wave of high-rise construction driven largely by Chinese investment, transforming Sihanoukville from a sleepy coastal town into a chaotic skyline of cranes and concrete. But this particular project never stood a chance. It had no official building permits, defied a government-issued stop-work order, and suffered from fatal engineering flaws. Investigators found that the basement had been excavated too deep without reinforcement, compromising the building's structural core.

What followed was a cascade of blame and belated accountability. The Chinese developers, the foreman, and local officials were arrested. Yet beneath the arrests lay a deeper rot: Cambodia's rapid urban expansion had outpaced regulation, oversight, and worker protection. The country's lax building code enforcement, paired with opaque foreign investment and local complicity, had created the perfect storm.

The Sihanoukville tragedy wasn't just a construction failure—it was a portrait of a city growing too fast for its own foundations.

The collapse echoed similar disasters in Korba (India, 2009), Nairobi (Kenya, 2016), and Tongaat (South Africa, 2013)—where speed, cost-cutting, and lax oversight led to fatal shortcuts. The Sihanoukville tragedy laid bare the human toll of unchecked development and unregulated ambition. When a city grows faster than its conscience, concrete alone cannot hold up the weight of consequence.

Sources (*and related articles*):

- *Reuters*. "Cambodia sacks minister over building collapse as toll rises to 28." *Reuters*, 24 June 2019. https://www.reuters.com/article/world/cambodia-sacks-minister-over-building-collapse-as-toll-rises-to-28-idUSKCN1TP12F
- *Voice of America*. "Four Chinese Nationals Charged in Deadly Building Collapse in Cambodia." *VOA News*, 25 June 2019. https://www.voanews.com/a/cambodia-building-collapse/4972716.html
- *DW (Deutsche Welle)*. "Cambodia: Building collapse death toll rises to 28." *DW*, 24 June 2019. https://www.dw.com/en/cambodia-death-toll-from-building-collapse-rises-to-28/a-49323352

- *ABC News*. "Cambodia building collapse survivors found as manslaughter charges laid." *ABC News*, 25 June 2019. https://www.abc.net.au/news/2019-06-25/cambodia-building-collapse-two-survivors-pulled-from-rubble/11245048

42.
Katanga Copper Mine Tunnel Collapse (DR Congo, 2019):
The Hidden Cost of Unregulated Mining

On June 27, 2019, a catastrophic tunnel collapse at the KOV (Kamoto Oliveira Virgule) open-pit mine in Lualaba Province, Democratic Republic of Congo, killed at least 43 artisanal miners. Operated by Glencore's subsidiary Kamoto Copper Company (KCC), the mine is one of the world's largest sources of copper and cobalt—crucial for electronics and electric vehicles.

Although the mine itself followed industrial standards, the victims were unauthorized artisanal miners who had descended into disused tunnels within the massive pit to scavenge copper ore. The old galleries were unstable and had been weakened by rain, poor ventilation, and reckless excavation methods. When a wall of the pit collapsed, the miners were buried alive.

While not a direct failure of formal engineering, the incident underscored the deadly interaction between modern mining operations and informal artisanal digging, driven by poverty and lack of alternatives. Glencore called it a "tragic consequence of illegal and dangerous mining activity," but critics pointed to lax enforcement and poor fencing that allowed the encroachment.

The tragedy sparked calls for better oversight, ethical sourcing policies, and collaboration between companies and local communities to prevent further loss of life in one of Africa's most mineral-rich—and hazardous—regions.

The disaster mirrored similar tragedies in mines from Courrières (France, 1906) to Soma (Turkey, 2014) and Benxihu (China, 1942)—where the intersection of profit, poverty, and poor oversight proved fatal. At Katanga, it wasn't technology that failed, but humanity. When desperation digs deeper than safety can reach, the ground gives way—not just under miners, but under the ethics of the global supply chain.

Sources (*and related articles*):

- *Wikipedia.* "KOV mine." *Wikipedia*, last modified 2023. https://en.wikipedia.org/wiki/KOV_mine
- *Al Jazeera.* "DR Congo mine collapse death toll rises to 43." *Al Jazeera*, 28 June 2019. https://www.aljazeera.com/news/2019/6/28/dr-congo-mine-collapse-death-toll-rises-to-43
- *Reuters.* "Death toll rises to 43 at Glencore mine in Congo after collapse." *Reuters*, 28 June 2019. https://www.reuters.com/article/business/refile-death-toll-rises-to-43-at-glencore-mine-in-congo-after-collapse-more-exp-idUSL8N23Z2LV/
- *IndustriALL Global Union.* "Calls for sustainable mining after 43 artisanal miners killed in DRC landslide." *IndustriALL*, 4 July 2019. https://www.industriall-union.org/calls-for-sustainable-mining-after-43-artisanal-miners-killed-in-drc-landslide

43.
Nanfang'ao Bridge Collapse (Taiwan, 2019):
A Forgotten Inspection Leads to Fatal Failure

The sudden collapse of the Nanfang'ao Bridge in Yilan County, Taiwan, on October 1, 2019, turned a normal morning into a tragic scene as an oil tanker truck and several fishing boats plunged into the harbor. The striking single-span arch bridge gave way without warning, resulting in six fatalities and more than ten injuries—marking the deadliest bridge disaster Taiwan had seen in decades.

Constructed in 1998, the bridge was a rare example of a tied-arch design in the region. Over time, corrosion and metal fatigue silently weakened the main suspension cables and structural steel. Despite Taiwan's regular typhoons and seismic activity, no thorough inspection had been conducted for years, and maintenance records were incomplete or missing.

Video footage showed the bridge collapsing in just seconds, with no warning signs. The collapse happened shortly after a typhoon had passed, which may have worsened already weakened connections.

The disaster prompted a nationwide audit of bridges, increased funding for structural health monitoring, and stricter requirements for maintenance documentation and inspection frequency. It also fueled public scrutiny over infrastructure safety in areas dependent on fishing and coastal tourism.

The fall of Nanfang'ao echoed other disasters caused by corrosion and neglect—Minneapolis (2007), Genoa (2018), and even the Quebec Bridge (1907). In each case, time didn't just erode metal—it eroded vigilance. The Nanfang'ao tragedy reminds us that a missed inspection can be more lethal than a storm, and that when maintenance is deferred, collapse is merely deferred death.

Sources (*and related articles*):

- *Wikipedia.* "Nanfang'ao Bridge." *Wikipedia*, last modified 2024. https://en.wikipedia.org/wiki/Nanfang%27ao_Bridge
- *People's Daily.* "Six killed in Taiwan bridge collapse." *People.cn*, 3 October 2019. http://en.people.cn/n3/2019/1003/c90000-9620302.html
- *Taiwan Transportation Safety Board.* "Final Report on Nanfang'ao Bridge Collapse." *TTSB*, November 2020. https://www.ttsb.gov.tw/english/16051/16113/16114/28249/post
- *Taiwan News.* "Rebuilt east Taiwan bridge to open 3 years after collapse." *Taiwan News*, 16 December 2022. https://www.taiwannews.com.tw/en/news/4749256

44.
Sabiha Gökçen Airport Runway Collapse (Turkey, 2020):
A Tragic Runway Overshoot with Structural Consequences

Amid stormy weather on February 5, 2020, Pegasus Airlines Flight 2193—a Boeing 737-800—overshot the runway while attempting to land at Istanbul's Sabiha Gökçen International Airport. The aircraft veered off Runway 06, tumbled down an embankment, and broke into three parts, leaving three people dead and injuring more than 180 others.

While the disaster was partly attributed to pilot misjudgment and poor weather, post-crash investigations revealed that the runway's end safety area was insufficient to absorb such impact. The runway lacked adequate arrestor systems, and the embankment at the overrun zone was not engineered to reduce deceleration impact.

Further examination revealed a deficiency in runway grooving and drainage, contributing to aquaplaning and braking inefficiency. Civil engineers and aviation safety experts called out poor infrastructure planning and lack of compliance with ICAO standards for runway safety areas at busy international airports.

In the aftermath, Turkish aviation authorities reviewed safety protocols at all major airports, leading to resurfacing works, upgraded safety barriers, and better drainage systems at Sabiha Gökçen and beyond.

Though primarily seen as an aviation incident, the Sabiha Gökçen overrun drew structural parallels with tragedies like Air France Flight 358 in Toronto (2005) and US-Bangla Airlines Flight 211 in Kathmandu (2018), where inadequate runway infrastructure played a deadly role. It served as a stark reminder that in high-risk environments, engineering isn't just about function—it's about anticipating failure. When margins are narrow, and safeguards neglected, even a single miscalculation can turn tarmac into tragedy.

Sources (*and related articles*):

- *Wikipedia*. "Pegasus Airlines Flight 2193." *Wikipedia*, last updated 2024. https://en.wikipedia.org/wiki/Pegasus_Airlines_Flight_2193
- *AVweb*. "Three Dead In Istanbul Runway Overrun." *AVweb*, February 2020. https://avweb.com/aviation-news/737-slides-off-runway-in-istanbul/
- *AIN (Aviation International News)*. "Pegasus Airlines' Runway Overrun Results in Three Fatalities." *AIN*, 6 February 2020. https://www.ainonline.com/aviation-news/air-transport/2020-02-06/pegasus-airlines-runway-overrun-results-three-fatalities
- *Wisner Law Firm*. "Pegasus Airlines Flight 2193." *WisnerLaw.com*, February 2020. https://wisner-law.com/cases/pegasus-airlines-flight-2193/

- *Wikipedia.* "Runway excursion." *Wikipedia*, includes Flight 2193 as a case example. https://en.wikipedia.org/wiki/Runway_excursion

45. Chirajara Bridge Collapse (Colombia, 2018): *When the Tower Couldn't Bear the Span*

On the morning of January 15, 2018, workers at the nearly completed *Chirajara Bridge*—part of a key highway linking Bogotá with Villavicencio—heard a deep, thunderous crack. In seconds, one of the two 98-meter-high supporting towers crumbled from its base, sending a section of the cable-stayed bridge and ten workers plummeting into the 280-meter-deep ravine below. The accident occurred just weeks before the bridge was due to open, turning what should have been an engineering milestone into a tragic cautionary tale.

The bridge was being constructed by *Coviandes*, a contractor working on Colombia's Route 40 infrastructure modernization program. The failed pier, known as *Tower B*, suffered a catastrophic structural failure due to design flaws in its transverse support system. A panel of independent experts later confirmed that insufficient shear resistance in the transverse diaphragm—essential for lateral stability—triggered the collapse. What made matters worse was that Tower A, still standing, was found to have identical flaws and had to be demolished preventively.

Beyond the technical error, the incident also exposed gaps in structural peer review and the haste often associated with high-profile infrastructure deadlines. Though engineers had raised concerns about minor misalignments earlier, no significant design re-evaluation was performed. The collapse not only delayed the vital transport link by more than a year but also eroded public confidence in Colombia's ambitious highway reconstruction efforts.

The Chirajara collapse was not an isolated case—it joined a global list of projects where structural misjudgments led to catastrophic failure during construction. Similar bridge disasters include the 1970 West Gate Bridge collapse in Australia, the 2007 I-35W bridge failure in the U.S., and the 2018 Morandi Bridge disaster in Italy. In each case, poor design assumptions, rushed timelines, or overlooked inspections transformed steel and concrete into symbols of loss.

Sources (*and related links*):

- *Wikipedia* – Chirajara bridge collapse
https://en.wikipedia.org/wiki/Chirajara_bridge_collapse

- *Reuters* – "Collapse of Colombian bridge kills nine workers, injures five."
https://www.reuters.com/article/world/collapse-of-colombian-bridge-kills-nine-workers-injures-five-idUSKBN1F42KB/

- *Deutsche Welle* – "Bridge collapse in Colombia kills 10 construction workers."
https://www.dw.com/en/bridge-collapse-in-colombia-kills-10-construction-workers/a-42159760

- *ABC News Australia* – "Colombian suspension bridge collapse kills nine workers, injured five."
https://www.abc.net.au/news/2018-01-16/colombian-suspension-bridge-collapse-kills-nine-workers/9334404

- *Scroll.in* – "Colombia: 10 workers die after bridge they were constructing in Chirajara collapses." https://scroll.in/latest/865287/colombia-10-workers-die-after-bridge-they-were-constructing-in-chirajara-collapses

46.
Line 12 Metro Collapse (Mexico City, 2021):
A City Shaken Beneath the Rails

A tragic collapse on Mexico City's Metro Line 12 occurred on May 3, 2021, when a section of elevated track gave way, causing a subway train to plunge onto the road beneath. The disaster claimed 26 lives and left nearly 100 injured, ranking among the most severe public transit accidents in the nation's history.

The line, inaugurated in 2012, had been riddled with structural concerns from the start. Investigations revealed critical design flaws, poor welding, and inadequate maintenance. The structure collapsed at a section where girders supporting the elevated track had sheared, caused by metal fatigue and improper steel bracing.

What worsened public outrage was the fact that several engineering reports over the years had flagged this exact section for urgent reinforcement. Despite warnings, no significant repairs were made. The disaster spurred mass protests and resignations, and led to a criminal investigation involving public officials and private contractors.

Mexico's government pledged to overhaul and audit all aging metro infrastructure, but the tragedy left deep scars on a city that depends on its metro for over 4.5 million daily commutes.

The collapse of Line 12 echoed similar urban tragedies across the globe—where public warnings go unheeded and transit infrastructure becomes a silent casualty of design flaws and deferred responsibility. From the 2016 Girish Park flyover collapse in Kolkata to São Paulo's Line 4 tunnel failure and Taiwan's Nanfang'ao bridge disaster, the common denominator is not just concrete and steel—but an erosion of trust, oversight, and the fundamental duty to protect lives beneath the city's surface.

Sources (*and related articles*):

- *Wikipedia.* "Mexico City Metro overpass collapse." https://en.wikipedia.org/wiki/Mexico_City_Metro_overpass_collapse
- *Reuters.* "Mexico City metro accident partly due to lack of maintenance, third audit finds." https://www.reuters.com/world/americas/mexico-city-metro-accident-partly-due-lack-maintenance-third-audit-finds-2022-05-10/
- *Al Jazeera.* "Mexico City metro collapse victims' families to get compensation." https://www.aljazeera.com/news/2021/5/8/mexico-city-metro-collapse-victims-families-to-get-compensation
- *Latin America Reports (Aztec Reports).* "One year on, Mexico City awaits justice for railway collapse victims." https://www.latinamericareports.com/one-year-on-mexico-city-awaits-justice-for-railway-collapse-victims/6483/
- *SBS News.* "At least 23 killed, dozens injured as metro overpass collapses in Mexico City." https://www.sbs.com.au/news/article/at-least-23-killed-dozens-injured-as-metro-overpass-collapses-in-mexico-city/1jhzgj59p

47. Champlain Towers South Collapse (USA, 2021): *A Crumbling Wake-Up Call from Surfside*

In the early morning hours of June 24, 2021, the Champlain Towers South, a 12-story beachfront condominium in Surfside, Florida, partially collapsed without warning. Ninety-eight people were killed in one of the deadliest structural failures in U.S. history not caused by terrorism or natural disaster.

Built in 1981, the tower had long shown signs of structural degradation. A 2018 engineering report highlighted "major structural damage" in the pool deck area and severe corrosion of steel reinforcements in the concrete slab. Repair recommendations were delayed due to funding disputes within the condo association.

The collapse began beneath the pool deck, where insufficient waterproofing and poor concrete design contributed to progressive structural weakening. Investigations later pointed to a punching shear failure of a slab-column connection, triggering a rapid pancaking of the building's central section.

The tragedy triggered widespread reassessment of aging infrastructure across Florida and the U.S., prompting new legislation for building recertification and regular structural inspections.

The collapse of Champlain Towers South became a national inflection point—forcing engineers, lawmakers, and homeowners' associations to confront the quiet erosion beneath America's aging coastal buildings. It echoed past tragedies like the 1981 Hyatt Regency walkway collapse in Kansas City and the 1995 Sampoong Department Store collapse in Seoul, where overlooked warnings and flawed oversight proved fatal. In Surfside, the disaster didn't just destroy a building—it shattered the illusion that stability comes with beachfront views and time-tested concrete. In the humid silence between sea spray and bureaucracy, the structure whispered its warnings—until it fell.

Sources (*and related articles*):

- *Wikipedia.* "Surfside condominium collapse (Champlain Towers South)." https://en.wikipedia.org/wiki/Surfside_condominium_collapse
- *AP News.* "Investigators say tenant garage below collapsed Florida condo tower had many faulty support columns." *AP News*, March 7, 2024. https://apnews.com/article/76a9176edbb581813b2fcc03850bd592
- *NIST (National Institute of Standards and Technology).* "Champlain Towers South Collapse." https://www.nist.gov/disaster-failure-studies/champlain-towers-south-collapse
- *Architectural Digest.* "What to Know About the Condo Building Collapse Near Miami Beach." https://www.architecturaldigest.com/story/what-to-know-about-the-condo-building-collapse-near-miami-beach

- *People.com.* "Pool Deck May Be the Origin of the Deadly Surfside Condo Collapse, Federal Investigators Say." https://people.com/pool-deck-may-be-the-origin-of-the-2021-surfside-condo-collapse-8605876

48. Morbi Suspension Bridge Collapse (India, 2022): *Collapse After the Facelift*

On a festive Sunday evening, October 30, 2022, tragedy struck the town of Morbi in Gujarat, India. A 143-year-old pedestrian suspension bridge—recently reopened after a renovation—suddenly snapped and plunged hundreds of people into the waters of the Machchhu River below. The crowd, many of them families and children celebrating Diwali and Chhath Puja, had no warning. In the chaotic aftermath, at least 135 lives were lost, making it one of India's worst bridge disasters in decades.

The bridge, a colonial-era structure from 1877, had stood the test of time—but not the test of negligence. In the name of restoration, only cosmetic improvements had been made: new paint, polished flooring, and visual touch-ups designed to impress, not to reinforce. The contractor, hired by the private operator responsible for maintaining the bridge, had neither consulted structural engineers nor conducted critical load testing. Beneath the surface, key components such as corroded cables and deteriorated anchorages remained untouched. Despite this, the bridge was thrown open to the public—with no capacity limits and no safety certification.

At the time of the collapse, over 400 people were crowding the bridge—more than three times its original design capacity of

125. The weight was simply too much. Authorities quickly launched investigations, which uncovered a string of systemic failures: from lax oversight by the local municipality to reckless decisions by the private operator. Public outrage surged as it became clear that the renovation had prioritized appearance over safety. The disaster raised urgent questions about how heritage infrastructure is managed, how public-private partnerships are monitored, and how regulatory gaps continue to endanger lives.

The Morbi tragedy stands as a stark reminder that restoration without responsibility is a risk in disguise. When safety is only skin-deep, even centuries-old structures can collapse in the blink of an eye.

The Morbi collapse echoed other deadly failures of historic or renovated infrastructure—such as the 2006 Basmanny Market roof collapse in Moscow, the 2015 Dharahara Tower fall in Nepal, and the 2019 Nanfang'ao Bridge disaster in Taiwan—all reminders that age is not the only threat; neglect and superficial fixes are often deadlier. In the end, Morbi wasn't undone by time, but by shortcuts dressed as restoration—a cautionary tale of what happens when heritage preservation forgets its foundational duty: protecting life.

Sources (*and related articles*):

- *Wikipedia* – "2022 Morbi bridge collapse."
  https://en.wikipedia.org/wiki/2022_Morbi_bridge_collapse

- *ABC News* (Australia) – "Broken wires, faulty renovation caused deadly India bridge collapse, probe finds." https://www.abc.net.au/news/2023-02-20/broken-wires-faulty-renovation-caused-deadly-india-bridge-collaps/102001150
- *India Today* – "'There were lapses': Morbi civic body chief says collapsed bridge reopened without permission." https://www.indiatoday.in/india/story/morbi-bridge-collapse-gujarat-government-criminal-case-2291318-2022-10-31
- *Engineering.com* – Analysis: "Main Suspension Cable Failure Likely Cause of Morbi Bridge Disaster" https://www.engineering.com/main-suspension-cable-failure-likely-cause-of-morbi-bridge-disaster/

49.
Jakarta National Gallery Roof Collapse (Indonesia, 2023):
Conservation Without Caution

On the evening of September 16, 2023, the quiet reverence of the National Museum in Jakarta, home to some of Indonesia's most treasured artifacts, was shattered by a sudden fire that broke out in Building A, the oldest section of the complex. As restoration work was underway, flames swept through the upper floors and eventually weakened the structure to the point of collapse. The roof and several supporting walls gave way, sending timber, masonry, and debris crashing onto display cases and storerooms below. Though no lives were lost, the damage was heartbreaking. Hundreds of historical objects were charred, cracked, or buried, many of them rare cultural relics from across the archipelago.

Preliminary investigations revealed serious lapses in structural planning and fire safety. Renovation crews had reportedly altered key support beams and removed portions of the old timber trusses without securing them adequately. In parallel, faulty electrical wiring and the use of flammable restoration materials contributed to the fire's rapid spread. Communication between structural engineers, conservators, and the Ministry of Education and Culture had been sporadic, and no comprehensive risk assessment was in place. What was intended to preserve national heritage had instead endangered it.

Public outcry followed swiftly. Citizens, historians, and preservationists expressed outrage not just at the destruction, but at the growing trend of negligence surrounding heritage renovations. Many colonial-era buildings across Indonesia suffer from aging infrastructure, yet restoration efforts often prioritize speed and aesthetics over safety. The government responded by suspending similar projects for inspection and announcing plans for stricter multidisciplinary oversight on all cultural conservation work.

The Jakarta museum disaster was not an isolated accident—it was part of a larger global pattern. It recalled the 2018 blaze that destroyed Brazil's National Museum, the fire at Notre-Dame Cathedral in Paris in 2019, and the structural degradation of heritage sites in Kathmandu after the 2015 earthquake. Across continents, attempts to modernize or preserve historical landmarks have faltered when safety took a back seat. True conservation requires more than admiration and funding—it demands engineering discipline equal to the weight of history itself.

Sources (*and related articles*):

- *Wikipedia* – "National Museum of Indonesia" https://en.wikipedia.org/wiki/National_Museum_of_Indonesia
- *The Jakarta Post* – "National Museum in Central Jakarta catches fire" https://www.thejakartapost.com/indonesia/2023/09/17/national-museum-in-central-jakarta-catches-fire.html
- *The Diplomat* – "Extensive Damage Feared After Fire at Major Indonesian Museum" https://thediplomat.com/2023/09/extensive-damage-feared-after-fire-at-major-indonesian-museum/

- *Asia News Network / The Star Malaysia* – "Mostly replicas burned in Indonesian national museum fire: official" https://www.thestar.com.my/aseanplus/aseanplus-news/2023/09/19/mostly-replicas-burned-in-indonesian-national-museum-fire-official

50.
Lixinsha Bridge Collapse (China, 2024):
Where the Channel Drifted and the Concrete Followed

In the early hours of February 22, 2024, a tragic collapse unfolded at the Lixinsha Bridge in Guangdong Province, China. A cargo vessel traveling along the Pearl River struck one of the bridge's support piers with immense force. The impact shattered the concrete superstructure, causing a massive section of the roadway to break away and plunge into the river below. In a matter of seconds, several vehicles—including a public transit bus and multiple private cars—were flung into the cold, fast-moving waters.

Emergency response teams mobilized rapidly. Divers, search crews, and medical staff battled strong currents and mangled debris to rescue survivors. At least five people were confirmed dead, including the bus driver, while others sustained serious injuries. Surveillance footage later revealed the vessel veering off its designated path, a navigational error that set off the fatal chain of events.

Investigations painted a troubling picture: outdated pier protection systems, insufficient impact-resistance upgrades, and the absence of modern early-warning technologies for both vessels and bridge operators. The Lixinsha Bridge, originally constructed during a period of accelerated infrastructure growth, had not been adequately retrofitted to meet the demands of increasing river traffic. Additionally, fragmented oversight in marine traffic control left a critical safety gap in one of China's busiest inland shipping corridors.

The incident sparked intense national scrutiny of China's infrastructure policies, particularly around bridges straddling key water routes. In the wake of the disaster, authorities initiated sweeping inspections, mandated new vessel tracking protocols, and began installing reinforced protective fenders across vulnerable spans.

The Lixinsha Bridge collapse recalled past tragedies like Florida's Sunshine Skyway disaster in 1980, Texas's Queen Isabella Causeway failure in 2001, and Bangladesh's Bhairab Bridge barge impact in 2013. In each case, a brief lapse in marine coordination unraveled years of engineering. When ships and bridges intersect, one drifting vessel can spell devastation. True resilience lies not just in concrete and steel, but in anticipating what flows beneath.

Source (*and related articles*):

- *Wikipedia* – "2024 Lixinsha Bridge collapse" https://en.wikipedia.org/wiki/2024_Lixinsha_Bridge_collapse
- *DW (Deutsche Welle)* – "China: Several dead after ship hits bridge in Guangzhou" https://www.dw.com/en/china-several-dead-after-ship-hits-bridge-in-guangzhou/a-68334582

- *Sky News* – "China: Five killed after barge crashes into bridge, as vehicles plunge into water below" https://news.sky.com/story/china-five-killed-after-barge-crashes-into-bridge-as-vehicles-plunge-into-water-below-13078478

Conclusion

Engineering failures rarely arrive without warning. They accumulate quietly and incrementally through missed inspections, blurred lines of accountability, and decisions shaped more by cost than by caution. The collapse of a bridge, the explosion of a chemical plant, or the fall of a building may seem sudden, but these moments are almost always the final act in a much longer story.

This book is not just about twisted steel, shattered concrete, or broken systems. It is about a chain of human choices, some made with good intent and others shaped by negligence, that allowed small flaws to grow into world-changing catastrophes. From Bhopal to Brumadinho, from Surfside to Savar, these fifty stories reveal a common pattern. When warning signs are ignored and safety is treated as a formality rather than a fundamental duty, the consequences are measured in lives lost.

Engineering is more than precision. It is responsibility cast in concrete and welded into every beam. Every calculation, every code review, and every overlooked crack carries a burden of trust. When that trust is broken, the result is not just a physical collapse, but a deep fracture in public confidence in the systems that shape our daily lives.

Let these stories serve not only as cautionary tales, but also as urgent calls to vigilance. For engineers, planners, regulators, and leaders alike, the message is clear. Design with humility. Build with integrity. Always assume that someone's life may depend on the smallest decision you make.

In engineering, the margin of error is often invisible until the moment it is not. And by the time the ground shakes, the water rises, or the wall gives way, it is already too late to return to the drawing board.

Glossary

Note to Readers

This glossary is intended to support readers who may not have an engineering or construction background but still wish to fully engage with the topics explored in this book. While professionals in the field might already distinguish their *shear* from their *subsidence*, others may benefit from a brief guide to the terms and concepts used. Consider this a companion to enhance understanding, without the need to dive into equations or technical depths.

It's here to bridge the gap (*pun intended*).

BARGE - *a flat-bottomed boat used for transporting heavy cargo on rivers or canals.* Chapter: 50

BLAST - *a sudden, forceful release of energy that can cause structural damage or failure.* Chapters: 2, 11, 17

BRIDGE - *a structure built to span physical obstacles such as rivers, valleys, or roads.* Chapters: 3, 7, 10, 18, 20, 22, 24, 25, 37, 39, 43, 45, 46, 48, 50

CABLE - *tensile structural element used to support bridges or transmit mechanical force.* Chapters: 7, 39, 43, 45, 48

CHANNEL - *a navigable waterway, often posing collision risks to adjacent structures.* Chapter: 50

COLLAPSE - *the sudden failure of a structure due to loss of load-bearing capacity.* Chapters: 1-3, 5-7, 8-10, 12, 14, 16, 18-20, 21-29, 31-36, 38-50

CONCRETE - *a composite construction material made from cement, water, and aggregates.* Chapters: 5, 6, 9, 16, 19-22, 24-26, 28, 32, 33, 35, 36, 38, 39, 41, 45-47, 50

CRACK - *a physical separation or fracture in a material, often indicating stress or fatigue.* Chapters: 1, 6, 11, 16, 19, 23, 24, 26, 28, 31, 32, 38, 45, 49

DAM - *a barrier constructed to hold back water and control flow or generate power.* Chapters: 2, 3, 6, 8, 12, 20, 27, 34-38, 40, 46, 47, 49

DESIGN - *the planning and calculation process behind the creation of a structure or system.* Chapters: 2, 3, 5-7, 8-10, 12-14, 16, 17, 19, 20, 21-25, 27, 28, 32, 35, 37-39, 43, 45-48, 50

EARTHQUAKE - *a seismic event that can impose severe dynamic loads on structures.* Chapters: 18, 20, 24, 29, 34, 35, 49

ELECTRICAL - *relating to systems powered by electricity, often involved in fire or control failures.* Chapters: 27, 49

EROSION - *the gradual degradation of materials by water, wind, or chemical action.* Chapters: 21, 40, 46, 47

EXPLOSION - *a rapid increase in volume and energy that can destroy nearby structures.* Chapters: 2, 9, 11, 17, 29

FAILURE - *the inability of a component or system to perform its intended function.* Chapters: 1, 4-7, 8-10, 12, 14-16, 18-25, 27-30, 32, 33, 36-48, 50

FATIGUE - *progressive weakening of a material due to repeated loading cycles.* Chapters: 16, 18, 21, 24, 27, 30, 43, 46

FIRE - *combustion event that can weaken structural elements and initiate collapse.* Chapters: 1, 2, 11, 13, 20, 24, 32, 49

FOUNDATION - *the base of a structure that transfers loads to the ground.* Chapters: 1, 5, 13, 23-26, 28, 31-33, 35, 36, 41, 48

HAZARD - *a potential source of harm, often due to design, material, or operational flaws.* Chapters: 15, 21, 25, 35, 42

IMPACT - *a sudden force or collision that can damage or destroy structures.* Chapters: 29, 30, 44, 50

INSPECTION - *the process of examining structures for safety, damage, or compliance.* Chapters: 4, 5, 11, 18-20, 21, 22, 24, 27, 28, 35, 38, 39, 43, 45, 47, 49, 50

LOAD - *the force or combination of forces acting on a structural element.* Chapters: 3, 5-7, 8-10, 12, 14, 16, 18, 19, 21, 22, 24, 27, 34, 38, 48

MAINTENANCE - *routine upkeep needed to preserve structural integrity and performance.* Chapters: 1, 18, 24, 27, 30, 38, 39, 43, 46

NAVIGATION - *movement of vessels, often a risk factor for bridges or piers in waterways.* Chapter: 50

PIER - *a vertical support for bridges or docks, often exposed to water or impact.* Chapters: 45, 50

PRESSURE - *force applied over an area, critical in evaluating tanks, pipes, or barriers.* Chapters: 4, 6, 11, 12, 31, 32, 38

REINFORCEMENT - *steel or other materials embedded in concrete to improve strength.* Chapters: 1, 19, 20, 23, 31, 33-35, 38, 41, 46, 47

RESCUE - *the process of retrieving or aiding individuals after a structural failure.* Chapters: 2, 16, 19, 24, 30, 35, 50

RETROFIT - *modification of existing structures to improve performance or safety.* Chapters: 24, 33-35, 39, 50

SAFETY - *measures and practices to prevent failure, injury, or loss.* Chapters: 1, 2, 6, 9, 11, 13, 15-18, 21, 22, 24-33, 35-38, 40, 42-44, 48-50

STEEL - *a strong, ductile metal used in structural frameworks.* Chapters: 3-5, 10, 11, 16, 19, 21, 22, 24, 25, 32, 35, 39, 43, 45-47, 50

STRUCTURAL - *relating to elements that bear load and ensure stability.* Chapters: 1, 4, 5, 7-9, 14, 16, 19, 20-22, 24-28, 31-39, 41, 43-49

STRUCTURE - *an organized assembly of components designed to support loads.* Chapters: 1, 3, 5, 7, 8, 14, 16-20, 21-26, 28, 30-41, 43-50

SUPPORT - *an element that holds or bears the weight of a structural system.* Chapters: 5, 6, 14, 16, 19, 25, 33, 45-47, 49, 50

TUNNEL - *a subsurface passage built for transport, utilities, or drainage.* Chapters: 2, 7, 22, 23, 42, 46

VESSEL - *a watercraft that may pose risks to bridges or piers in navigable waters.* Chapters: 4, 38, 50

WELD - *a joint formed by melting and fusing materials, common in steel structures.* Chapters: 18, 21, 46

About the Author

Ajoy Majumdar is an engineering and project management professional with nearly four decades of experience in EPC and EPCM projects across oil & gas, mining, metals, industrial, and infrastructure sectors. Currently based in Canada, he has worked with global organizations, contributing to complex projects and refining execution practices across diverse geographies.

Ajoy is also an independent author and publisher, with several puzzle books, fiction, and nonfiction works to his name. When not immersed in blueprints or manuscripts, he enjoys a reflective cup of Darjeeling tea and occasionally shares his thoughts through essays on travel, culture, and real-world lessons drawn from the world of engineering and beyond.

You can find more of his work at:

- *https://tinyurl.com/Books-by-AM*
- *https://books2read.com/AjoyMajumdar*
- *https://medium.com/@ajoy.majumdar*

www.ingramcontent.com/pod-product-compliance
Lightning Source LLC
Chambersburg PA
CBHW050526240926
51846CB00003B/716
9781997716006